I0492382

# The *TRUE* Real Story of Creation

## T W Manes

## The *TRUE* Real Story of Creation

Copyright 2020 © by T W Manes

All rights reserved.

No portion of this book may be used or reproduced in any manner whatsoever without the author's expressed written permission, except in cases of brief quotations embodied in critical articles and reviews. For permission, contact **tw502k@gmail.com**

All scripture references are from the King James Version unless otherwise noted.

# INTRODUCTION

T he story of Creation as told in hundreds, if not thousands, of other books and articles is as varied, chaotic, and incongruent as the supposed difference between science and religion is. So, this book you are about to read is my attempt to combine most, if not all, of these differences into an easy to read, easy to understand, and especially, easy to *believe* book, no matter which side of the story of Creation spectrum you fall on.

So, without further ado, how about if we start this book off with an undeniable absolutely true fact concerning Creation, that without a doubt, no matter what your beliefs are about Creation, everyone can all believe in?

Is that okay with you?

Well, believe it or not, there really is one of those very rare items. So here goes; that undeniable fact is that Creation really and truly actually took place!

Want proof?

Look in a mirror.

Yeah! The proof is that you and I actually exist, and thus, that my friends, proves that Creation actually happened, *somewhere back in time!*

But following that huge acknowledgement, how can we know just what really is, "The *True* Real Story of Creation"?

Is the Scientific exposition of events the correct, and *only*, version of truthful 'facts', or is the fantastical version of events written in the Hebrew and Christian Bibles the correct, and *only*, version of truthful 'facts'?

Or, is there a way by some possible supernatural coincidence, or providence, a place where the two wildly differing versions of events can be combined to reveal the real, undeniable *truth* about Creation?

Well, my sincere belief is that by the time you finish reading this book, there will be so many undeniable true *facts* pulled out of 'both' versions of the Creation story that you will never have any more doubts about "The True *Real* Story of Creation" ever again.

But as the wise old sage who sits on top of the mountain once said; "Grasshopper, there are many obstacles in the way to overcome before you can get there."

Now that, too, is an absolute true 'fact'!

But first, my friends, please allow me to take just a few minutes to introduce myself to you and to give you some reasons just why I decided to throw my hat into this hugely divided ring of controversy concerning Creation.

It's very simple; most of the time when the scientific versions of Creation have been written, they have usually been presented in such a way by the authors as not trying to really educate us about Creation, but more so an attempt to prove just how much smarter and how much more intelligent they think they are than everyone else is! I mean, most of the time, I can hardly make myself continue reading after getting no more than halfway through a paragraph, or even a sentence. In actuality, usually I just get tired of having to look up the 'big words' they keep using in the dictionary.

I mean, come on, man! Can't you scientifically possessed people just try to get your points across in a manner most of us *'normal'* people can read and understand?

And as for most of the religious leaders?

I mean, come on, man! Can't you just try to get your points across without being so fanatically dogmatic? I mean, for most of us church going people who may have the inclination to become members of your church or denominations, one very troubling requirement for many of us that we first have to do, is we have to verbally swear our confirmation and agreement in your proclaimed dogmatic denominational statement that every word written in the Bible is *infallible and inerrant in all of its aspects.*

So, maybe I'm being foolish here, but I truly believe that I can actually write a book about the Scientific version of Creation whereby most all people can actually understand it, and actually 'enjoy' reading it at the same time.

And, maybe I'm being foolish here, again, but I truly believe that I can include in this book the Biblical version of Creation whereby most all people can understand and actually 'enjoy' it, too. The reason I say this is because I, personally, am a huge believer in both versions of Creation.

But you may ask, 'how can that possible be? The vast chasm between these two versions is just too far apart to ever be combined together into a cohesive believable story.'

Well, if I didn't believe that it is actually possible, then why else would I write a book with the absurd title of, *"The TRUE real story of Creation"?* I mean, do you actually think that I am so mentally deficient that I willingly want to risk the wrath of the intolerable Cancel Culture by writing this book?

No way, man! So, again, as I began to walk down this seemingly deep, dark, bottomless crevasse of illogical reconciliation, the wise old Sage's advice about there being many obstacles in the way proved completely correct. You see, before I even began putting pen to paper, or more precisely, chubby stiff fingers to a keyboard, I was studying everything I possibly could about Creation and History, and about how Mathematics has played such a big part in it all.

But, at the same time, I was having what could generously be called, a huge conflict of interest. You see, from my earliest memories, I had been taken to church. And as most of you can imagine, or possibly even attest to yourselves, what I was taught there concerning the story of Creation and History quickly led me on a direct collision path between Science and Religion where I actually thought there was absolutely no chance of reconciliation between the two versions. I actually began to believe that the obstacles the wise old sage had said would present themselves were just too great to be overcome!

But you know what I eventually learned?

I learned that just about anything is possible if you just put enough time, effort, and firm belief into it! The obstacles can be overcome, no matter how wide and deep the chasm is. So, what once looked completely incompatible to me, a firm path of reconciliation between Science and the Bible has now not only become possible, but a proven 'fact'.

And I think you will not only be surprised, but may even have your socks blown right off your feet by the results. My quest to become the new, old wise sage is now fulfilled. But, the obstacles in my way to present this knowledge to you, Grasshopper, are many.

One huge obstacle is, how can I impart this information to you in such a way as to keep from boring you so badly that I will be guilty of putting most, if not all, of you into an early eternal Big Sleep like most of the previous writers of the Science version of Creation?

So please allow me to put in a huge disclaimer right here. As a necessity, this book must have both, an abundance of scientific facts, along with a slew of biblical scriptural references, included in it. But my friends, I promise that I will do everything possible to keep you interested, awake, and most of all, entertained!

What is the purpose of writing a book if your readers are going to dose off from absolute boredom?

So, I believe that you will find the following True *Real* Story of Creation a fun, enjoyable, and especially, an entertainingly written history of the story of Creation that is guaranteed to keep you awake no matter which side of the debate spectrum you fall on.

So, as we begin, let me ask you a question. Do you believe that the story of Creation as written in the Hebrew and Christian Bible is such a seriously flawed fantasy tale that is full of so many holes that it looks like my 20-year old favorite

T-shirt? Or do you believe that the Creation story as told by Science is so blasphemous that there is a special place already reserved in Hell for those evil atheist apostates?

Well, for the first time ever, 'someone' (me, lol) has dared to travel down the meandering stumbling drunken crooked path that flows fluidly between fanciful fantasy and indisputable fact, and has separated and combined them to the Nth degree to finally tell the *"true real" story of Creation!* So, here's your chance to be amazed and informed as every question you have ever had about the *when's, where's, why's, and how's* is finally answered in a coherent, and especially, an undeniable believable fashion!

**A small sampling of some of the questions answered:**

Did a Supreme Intelligent Being, who is infinitely more intelligent than any of us mere doleful mortals, really spit a speck of Light out of His mouth that actually brought about the creation of the Universe and everything in it? Or did some kind of an unknown indeterminate Cosmic Happenstance suddenly burp Creation into existence?

Well, whatever caused it, Creation did take place! And that folks is an absolute _TRUTHFUL FACT!_

So, did Creation really only take just _6-days_ as the fervent Bible-thumping religious Righteous believe? Or is the true age of the Universe much closer to the _13.799-Billion years_ as the fervently believing Believers in Science declare?

Also, Science states that the Earth was really created a little more than _4.5-Billion years_ ago, while the Bible says that the Earth was created on the _3rd day_ of Creation, which according to many theologians and fervently believing religious Righteous; believe this happened somewhere only around 7,000 to 10,000 thousand years ago.

Who, or which, is true? Or can it even be factually determined?

Was "Light" really what created the Universe? If so, then where did the Light come from? And, more importantly, how could that first burst of _'Light'_ contain _'everything'_ inside of it that has _ever,_ or _ever will,_ exist in the Universe?

Listen, friends, there are hundreds, if not

hundreds of thousands, of questions concerning *the true real story of Creation.* But to attempt to list them all here would take way too much time, and the list would be such an impossibly huge list that it would, without a doubt, bore the unholy living crap out of you that would probably put the majority of you into a permanent catatonic state of the Big Sleep! Besides, most of these questions will be answered within the context of this book as we go forward.

So just what is the truth? And just how can this seemingly insurmountable and astronomically huge conundrum of facts and fiction ever be reconciled to give a *'true real'* story of Creation?

Well, it can be done. I promise you that in the following pages are enough scientific *facts* to keep an unimpressible atheist happy and completely satisfied! As well as there being enough actual *truths* from the Biblical perspective to make most Christians happy and satisfied as well!

All I ask is that whichever 'group' of believing Believers that you fall into, you will happily and studiously go forward with a complete, uncluttered, and open mind, truly and honestly willing to accept whatever conclusions the following *truths and facts* reach, and if you will, I guarantee that both groups will be truly amazed, utterly astounded, and in the process, totally entertained!

But for this to happen, all of you uncompromising and unmovable, fanatically inspired, die-hard believing Believers in the scintillating Scientific version of Creation must open your minds to be willing to try to understand just why, the uncompromising and unmovable, fervently religious, fanatically inspired, Bible-thumping believing Believers are so adamant about their beliefs, and of course, *vice versa!*

And as for you uncompromising and unmovable, die-hard, fervently religious, fanatically inspired, Bible-thumping believing Believers, since you believe without any deviation in the absolute inerrancy and infallibility of the Holy Scriptures, you then, are also uncompromisingly forced to believe whole-heartedly in the following scripture: *2 Timothy 2:15 "Study to shew thyself approved unto God, a workman that needeth not to be ashamed, <u>rightly dividing the word of truth."</u>*

Why do I quote this scripture?

For several reasons. But the main one is that the vast majority of Christian Believers have based almost all of their uncompromising religious beliefs wholly on whatever their pastors, Sunday School teachers, priests, or any other esteemed figures in religious authority, have told them to believe, or they would be eternally damned to the furthest reaches of a fiery eternal Hell if they didn't.

So, I dare to ask those of you who fall into this category; how many of you have ever really done an in-depth study of the Scriptures concerning Creation to see for your own self if what you've been taught is actually true, *or not?*

And if you have really, and honestly, done this, I must then ask just how many of you have actually expended the time, energy, and endurance to go all the way back and compare the version of the Scriptures that you carry in your hand to church each Sabbath with the *'original'* Scriptures the Holy Books were originally written in?

For example: What Christians call the Old Testament was originally written in the *Hebraic language.* So, I must ask, how many of you have actually compared the beautifully written version of Hebrew Holy Books, to say, the King James Version, since the King James Version was the first version to be translated into the English language?

Well? . . . I'm waiting!

Yeah! That's what I thought.

Not too many of you spoke up. So, with that being said, I insist that you open your closed Pharisaical minds to the possibility that you just may learn a brand-new version of how, why, when, and where Creation really began, and how it has continued down throughout all history.

In other words, do what the above scripture says you must do if you want to > *Make God proud of you by being a true 'workman' who needs not be ashamed, because, you have actually taken the time, expended the energy, and exuded the endurance, to really and truly, study the scriptures (word), and then, to 'rightly divide them until you really and truly, reach the unyielding, TRUTH!!!*

**Note:** Later on, in this tome, I am going to be expounding on this theme in much greater detail. So please pay attention. *You will be tested on it.* ☺ (Not really)

And as for you total, uncompromising, die-hard believing Believers in the Scientific version of Creation, get the smirk off your face because, I guarantee you that in the following pages, all of your closed-minded, non-triggered beliefs are about to smack you upside your head so hard that undoubtedly, you are going to have such a gigantic migraine that just won't go away until you're willing to possibly understand why those so-called silly Christians believe the way they do!

So, with all of that being said, let us begin.

# CHAPTER 1

## "In the Beginning"
## The Creation Timeline
## 13-Plus Billion Years Ago, or 6 - 7
## 'Days' Ago

**S**ince Creation is an undeniable fact that I truly believe we can all agree on without much dissent, especially since we're all here right now as pure living proof, that leaves the one huge, major, most astronomically important question of all: *When did Creation actually take place?*

Six-days ago?

Or a really, really, $r - e - a - l - l - y$ long time ago?

Well, it is actually impossible to give a true real definitive time to the *actual exact instant* that "LIGHT" impossibly somehow suddenly appeared out of what was complete *nothing,* and then quickly thereafter, that Light suddenly blasted forth out into the vast empty 'void' of total and complete *'nothing'*.

Seems like *'NOTHING'* is kind of important here, huh?

Well yeah! If you totally believe in the scientific version of Creation that before Light somehow appeared, *nothing* was the only *thing* (if *nothing* can even be described as a *'thing'*) to exist at that time.

Anyway, the closest estimate that Science has been able to give to this date *at this time,* is that around *13.799 billion years ago,* a form of *LIGHT* suddenly blasted outwards from some kind of an unknown source into the vast *empty* surrounding void *in all directions!*

So, to answer another question presented earlier; if you want to know *"where"* Creation actually began, the answer is that it was, and still is, located directly in the very *'center'* of our Universe. That is because of the undeniable fact that the 'light' burst outwards *in all directions!*

So, if mankind were ever able to create 'something' that could actually find the location of the very center of the Universe so they could peer directly into it, there they would find the glow from the very first remnants of Light to ever exist. Or to those of you who believe in a place called "Heaven", it is very probable that it is located right there in the very 'center' of *everything* that now exists!

***This fact is very important!*** Please remember this as we go further into this story of Creation!

Thus, that magnificent, wondrous, and completely indescribable *miraculous* explosive beginning of Creation is when *time*, as mankind has determined and interpreted it to be so, first began. And for the complete lack of any better intelligent ability that could possibly give this gigantic, unimaginable, and completely indescribable explosive energy of *light* some kind of a better name, mankind has simply labeled the sudden vast explosion into the vast empty void, *"The Big Bang"!*

So, my friends, this is the scientific explanation of when and where Creation took place. But, is this *the true real story of Creation?*

We'll soon determine that with a very undeniable definitive answer! But first, you must allow me to be fair to the religiously fervent, Bible-thumping believing Believers by allowing me to present their belief of how the true real story of Creation took place. The following is the King James Version of the Creation story of the Universe:

## Genesis Chapter 1

**1 In the beginning God created the heaven and the earth. <sup>2</sup> And the earth was without form, and void; and darkness was upon the face of the deep. And the Spirit of God moved upon the face of the waters. <sup>3</sup> And God said, Let there be light: and there was light. <sup>4</sup> And God saw the light, that it was good: and God divided the light from the darkness. <sup>5</sup> And God called the light Day, and the darkness he called Night. <u>And the evening and the morning were the first day.</u>**

**<sup>6</sup> And God said, Let there be a firmament in the midst of the waters, and let it divide the waters from the waters. <sup>7</sup> And God made the firmament, and divided the waters which were under the firmament from the waters which were above the firmament: and it was so. <sup>8</sup> And God called the firmament Heaven. <u>And the evening and the morning were the second day.</u>**

*⁹ And God said, Let the waters under the heaven be gathered together unto one place, and let the dry land appear: and it was so. ¹⁰ And God called the dry land Earth; and the gathering together of the waters called he Seas: and God saw that it was good. ¹¹ And God said, Let the earth bring forth grass, the herb yielding seed, and the fruit tree yielding fruit after his kind, whose seed is in itself, upon the earth: and it was so. ¹² And the earth brought forth grass, and herb yielding seed after his kind, and the tree yielding fruit, whose seed was in itself, after his kind: and God saw that it was good. ¹³ <u>And the evening and the morning were the third day.</u>*

*¹⁴ And God said, Let there be lights in the firmament of the heaven to divide the day from the night; and let them be for signs, and for seasons, and for days, and years: ¹⁵ And let them be for lights in the firmament of the heaven to give light upon the earth: and it was so. ¹⁶ And God made two great lights; the greater light to rule the day, and the lesser light to rule the night: he made the stars also. ¹⁷*

*And God set them in the firmament of the heaven to give light upon the earth, ¹⁸ And to rule over the day and over the night, and to divide the light from the darkness: and God saw* that it was good. [19] *And the evening and the morning were the fourth day.*

[20] *And God said, Let the waters bring forth abundantly the moving creature that hath life, and fowl that may fly above the earth in the open firmament of heaven.* [21] *And God created great whales, and every living creature that moveth, which the waters brought forth abundantly, after their kind, and every winged fowl after his kind: and God saw that it was good.* [22] *And God blessed them, saying, Be fruitful, and multiply, and fill the waters in the seas, and let fowl multiply in the earth.* [23] *And the evening and the morning were the fifth day.*

[24] *And God said, Let the earth bring forth the living creature after his kind, cattle, and creeping thing, and beast of the earth after his kind: and it was so.* [25] *And God made the beast of the earth after his kind, and cattle after their kind, and everything that creepeth upon the earth after his kind: and God saw that it was good.*

[26] *And God said, Let us make man in our image, after our likeness: and let them have dominion over the fish of the sea, and over the fowl of the air, and over the cattle, and over all the earth,*

*and over every creeping thing that creepeth upon the earth.* <sup>27</sup> *So God created man in his own image, in the image of God created he him; male and female created he them.* <sup>28</sup> *And God blessed them, and God said unto them, Be fruitful, and multiply, and replenish the earth, and subdue it: and have dominion over the fish of the sea, and over the fowl of the air, and over every living thing that moveth upon the earth.*

<sup>29</sup> *And God said, Behold, I have given you every herb bearing seed, which is upon the face of all the earth, and every tree, in the which is the fruit of a tree yielding seed; to you it shall be for meat.* <sup>30</sup> *And to every beast of the earth, and to every fowl of the air, and to everything that creepeth upon the earth, wherein there is life, I have given every green herb for meat: and it was so.* <sup>31</sup> *And God saw everything that he had made, and, behold, it was very good. <u>And the evening and the morning were the sixth day.</u>*

---

So, the question must be asked, is *this* description about taking only six "days" for God to create everything *the true real story of Creation?*

---

Well, to be perfectly honest, as I was growing up, I was taught this version so fervently as a true actual undeniable fact that any unbelief in it would have condemned me to a dark, dank, and fiery eternal hell. But, even so, it has always been a huge problem for me to believe. I could just never get past this "six days" alleged *fantasy*, as I believed!

So, eventually, I fell into the trap of believing that the *order of creation* listed in this version is, at least, pretty close to being correct but, the timeline of just six days for all of Creation to be completed was just a mad version of an insane man named Moses who is credited with writing this story whose mind had been utterly scrambled by too many hard and long years roaming around in the blazing hot sun of the Arabian Desert keeping watch over a herd of hungry and thirsty sheep!

But, deep down inside, I knew that could not be correct, either, because of the fact that it was Moses that was the person who wrote these Scriptures. You see, Moses lived somewhere around three-thousand five-hundred years ago. But yet, he still somehow inexplicably knew that *"Light"* was the first thing that ever existed in Creation, which is confirmed by Science?

Now just how could that be?

And, he somehow inexplicably also knew that the first *living creatures* to ever appear on the Earth originated in 'water'.

Again, how could he have known that?

So, the conundrum I had was how could a man that wrote these Scriptures somewhere around three-thousand five-hundred years ago know everything that he knew if it wasn't correct?

So, I must ask you the same question. Just how could it have been possible for a man who lived a way, way long time ago, and who had spent 40 years of his life living in the blistering desert as a sheep herder know that *Light* was the first thing ever Created in history? And, especially, since he was a sheep herder living in the middle of a huge desert, just how in the world could he have possibly known about the first life to originate in the world was in water?

"But, yeah", you say, "maybe he just guessed about the Light and life thingy, but come on man, we all know that it took way more than just *six days* for all of Creation to happen! Anyone with any commonsense at all would know that!

Even the 'flat-earth' people know that it took a whole lot more than six days, and they have the least commonsense of anyone at all."

I will get into this 'six days' thingy, as you put it, in just a minute. But first, I just have to ask you; do you really think an eighty-year-old sheep herder could have just guessed that *'light'* was what burped Creation into existence? Or that the same eighty-year-old man living in the middle of a vast desert somehow knew that *life,* that began on Earth several billion years before him, began in the waters of the world?

So, I must say that your answer of Moses maybe just guessing about the light and life thingy holds no possible heartbeat nor any water. *Pun intended!*

But I must ask you this, if you were to look deep inside of your heart, is it even a smidgen of a possibility that just maybe the answer of how Moses knew these things is what is written in his book called, Genesis? You see, written right there in Genesis chapter 1, verse 4, Moses wrote these words, *"And God said".*

I won't ask you to answer that question at this time. But please keep it in the back of your mind.

Now, let's get back to your other bold statement about it being impossible for the Earth to have been made in just six days. I absolutely see your point. And I must say that I absolutely agree with your statement, too.

But what if I told you that I finally realized the answer to this seemingly impossible conundrum and that the answer has a really simple solution?

"Okay, sure man. Go! Make a believer out of me! He he he!!!"

Okay, I will. So, listen up.

First of all, I want you to take a very intensive look at the above Scriptures again. Now, tell me this, what language are they written in?

"English, man. Anyone can see that."

Oh, you're so good, my friend. I compliment you. You're really cognizant. But you know something; to what will assuredly be to the total surprise, and possible dismay of many people, including a tremendous number of biblical teachers and even theologians, English was not the *'actual'* language that Moses used to write those Scriptures in.

You see, unfortunately, many Christian denominations believe that the *English* version of the Bible, usually the King James Version since it was the first version to be translated into English, is so fool-proof that they claim it is 'inerrant' and 'infallible' in all of its aspects. In fact, in many churches and denominations, they even make you affirm that belief either in writing, or at least, verbally, before you can become a member of their denomination or church. But what they have apparently never taken into consideration, or more than likely, 'hid' from you, is that the people who translated the Bible into English were not the *original authors* of any of the Books of the Bible!

"Huh?"

Yeah! You see, the religiously fervent Bible-thumping believing Believers believe that every word written in the Bible was truly inspired by God, and every one of those words is completely inerrant and infallible in all of its aspects.

So, here's where the mouth bites the bullet, or something like that.

As we've already determined, Moses was the writer of the story of Creation in the book of Genesis. ~ Actually, he was the author of the first five books of the Hebrew Bible, which is called the _'Torah'_, along with the first five books listed in the Christian _'Old Testament'_. These 'books' are one and the same. ~ Anyway, here's the thing; Moses was a "Hebrew" man. Thus, surprisingly, (or not) when he wrote the above Scriptures, he actually, really and truly, used the Hebrew language to do so.

So, let's follow this new amazing thread of pure brilliant deductive knowledge one more step to its indisputable logical conclusion. First of all, please allow me to say this; much of the Hebrew vocabulary ~ same as English ~ can have multiple meanings for the same words depending in what _context_ the words are put into. So, to begin our journey, let's do a little investigation and just see what the _Hebrew word_ translated for _'day'_ in the above _'English'_ Scriptures actually is, and then let's see for the sake of sanity, if that Hebrew word has any other meanings than the word _'day'_.

The Hebrew word translated day is _"yom"_. *Wikipedia*

Now, for some ridiculous reason (my own hypothesis) <u>*yom*</u> has been translated as *day* in the English translations of the Bible, but the word <u>*yom*</u> actually has several other definitions as well: It can also be interpreted as:

- ***Period of light*** (as contrasted with the period of darkness),
- ***General term for time***
- ***Point of time***
- Sunrise to sunset
- Sunset to next sunset
- **A year**  *(in the plural; I Sam 27:7; Ex 13:10, etc.)*
- <u>***Time period of unspecified length.***</u>
- <u>***A long, but finite span of time- age - epoch - season.***</u>

**S**o *okaaaayyyy*, looking at the above actual different meanings of <u>*"yom",*</u> it makes one wonder just why the translators of the Hebrew Bible into English used the word *'day'* to describe the timeline of the story of Creation! Were they just plain lazy and didn't want to take the time to look, or *think*, just a little further? I mean, *six* out of the *eight* meanings of *yom* give a definite meaning far different from the word 'day'. One of them translates to a 'year'. Even that meaning is still ridiculously incorrect in the time context of Creation, but a 'year' is still a much better translation than 'day'.

So, I just have to ask, is this just unbelievable?

Well, yeah!

And that's why so many people ridicule the Biblical story of Creation in the Bible. And they have a great point, too. Just take a minute to think about how gloriously different things might have been, and especially, how much more relevant and believable the Bible would have been received, if the *translators* had used *'a time period of unspecified length' or, 'a long, but finite span of time - age - epoch - season'* to describe the sequences of events!

Listen, I can almost guarantee you that when Moses was writing these Scriptures, he wasn't thinking that Creation took only six *actual* 24-hours a day to complete! Moses was not ignorant, nor was he delusional! He was a very intelligent and educated man, having spent the first forty years of his life growing up in Egypt as the adopted grandson of the Pharaoh. As such, he had been educated in the best, most advanced studies known to mankind at that time.

In fact, later on in his writings, he wrote the Laws that still form the basis for the legal system for the nation of Israel even to this day! And, I must also emphatically say, he also wrote the ten most important *moral laws* to ever be written, too!

So, it is just downright inconceivable that Moses would write something so utterly nonsensical about the timeline of Creation that it's truly beyond the realm of reason!

Also, I truly believe that since Moses was educated in Egypt, and since their form of writing at the time was Hieroglyphics, or *picture writing*, (word pictures) that when Moses was writing the story of Creation as inspired by God, he tried his best to tickle his current audience's imaginations with *word pictures* in a way that these uneducated former slaves would be able to best understand. These former slaves knew what a *'day'* was, because they worked from sun up to sun down every day. But if Moses had tried to tell them that Creation first took place over *13-billion* years ago, they would have probably stoned him, because their minds just could not conceive of anything like that.

I ask, can your educated mind conceive of that length of time in a rational way even now? Or mine, either?

Even so, I believe I know why the *translators* used the word "day" to describe the Creation story. And no, I don't believe they were just plain lazy, either. I just think they were confused by the way Moses described Creation.

You see, by remembering the words of 2nd Timothy 2:15, and by *'rightly dividing'* the first verses of scripture in Genesis, I believe that we have to take the words in the "context" Moses was using them in. But in addition to that, we also have to understand the scriptures in the *context* of who his *'audience'* was at the time he wrote those words. So, I believe that what Moses was actually doing was breaking Creation down into *six different parts.* I believe this was so his 'audience' could understand just what in Heaven's glories he was describing to them. So, in order to break it down for them to be able to understand, he used the terminology; "the **evening** and the **morning** was the "*day*".

Now, I have always wondered why Moses described a so-called *'day'* by stating it as an *evening,* then a *morning.* That's not a *day* to the average person. That would be a *'night'*. Also, at the very most, it would only be a maximum of 12-hours, or there about.

So, my logical conclusion was several things. First, the 'evening' is usually the time when a 'day' (or a 'period of time') is winding down and coming to a close, and will be that way until the next sunrise. So, my belief is that Moses was just trying to simplify Creation *to his 'simple' people* in the best way he could by explaining that each portion of Creation took only a *day* of **God's 'span of time'**. And I believe he was using this *'span of time'* by God as a comparison to their 12-hour work days they were used to as slaves!

But, of course, I'm sure that it never entered into his thinking about what a mess it would cause 3500 years later on to a new, and mostly, highly intelligent audience!

And another reason, I believe, was that he wanted to emphasize very strongly that these former slaves no longer had to work seven 'days' a week anymore without rest like they had to do under their former masters. He wanted them to take every seventh day and rest on that day, like he had described God as doing after God completed His work. But, he wanted for them to also, spend that 'day' in remembrance and by giving thanks to Almighty God, *Yahweh,* their Creator.

So, listen, when Science tells us that approximately 13.799-billion years ago the Big Bang took place; please feel free to believe that. I certainly do, just like Moses, the 'original' *inspired* author of the book of Genesis did!

So, to all of you religiously fervent, Bible-thumping believing Believers, now maybe you can understand why the Apostle Paul wrote the following scripture; ***"Study to shew thyself approved unto God, a workman that needeth not to be ashamed, rightly dividing the word of truth."***

If you don't study as a fervent workman in such an in-depth manner to be approved by God, then you will never be able to *rightly divide the word(s) of truth!* As proven by the fact that the word 'day' has been allowed to remain in the English translation versions of the Bible without reproof.

Listen, I firmly believe that the *original authors* of the various books of the Bible were without a doubt, truly "inspired by God" to write them. But does that mean that the _translators_ that translated the books were just as inspired? Or could it have just been a paid job for them, such as was the case of many of the books included in the New Testament.

Bet that one took you by surprise, huh? Well, it's your fault! Study to show yourself approved next time!

Listen, you can absolutely have faith that every book; in fact, every 'word', written in the _Hebrew Bible_ has been copied exactly correct from copy to copy going all the way back to the original authors of those books because each of those books were transcribed only by Hebrew Priests, and if these priests made even the tiniest mistake in any way possible, the entire scroll was then taken out and burned! That's how important they took the words they believed wholeheartedly were totally inspired by God.

But as for the New Testament books, every time a copy was sent to a new location where a new congregation had risen up, another copy of the book was transcribed for that congregation to keep in their possession. Now, if no one there knew how to read and/or write, then a paid scribe was employed to copy the text. And, unfortunately, in some cases, words and phrases that were *not* inspired by God were included in the books.

You can realize this if you properly *rightly divide the words of truth* in the books written by Paul. In some of his books, he forcefully declares that 'women' are not to be allowed to speak in the church, and therefore, they certainly could never be any sort of a leader or teacher in it. *(Let your women keep silence in the churches: for it is not permitted unto them to speak; but they are commanded to be under obedience, as also saith the law. 1 Corinthians 14:34)*

But in another of Paul's books, in **Romans 16:1-2** he declares that he is working alongside 'Phoebe', who he calls a 'Deaconess' in the church; or, a 'leader' in the church. **(*Now I commend to you our sister Phoebe, a <u>deaconess</u> in the church at Cenchrea.*) *New International Version (NIV)***

And, there are several other glaring *contradictions* in his books if you *read, "study",* and then, *"rightly divide"* them closely.

I am absolutely, one-hundred percent positive that these glaring contradictions were not 'originally' written by Paul! Because, just as in the Old Testament books, I firmly believe that Paul, along with all of the other New Testament authors, were one-hundred percent inspired by God to write their *'original'* books! But, unfortunately, the tight dogged oversight of the Jewish Priesthood in making absolutely sure that no additions or subtraction were done to any of their Holy Books was not done by the different new churches that had sprung up in the early 'Christian' era.

Without a doubt, certain words and verses were added, and even subtracted, by 'inspired' scribes of Satan such as the one who had a thing against women, and added the verses in Paul's writing about women must be silent at all times in church!

Want proof?

Compare those so-called 'scriptures' supposedly from Paul degrading women to what Mohammad had transcribed later on in the 8th century! But being that Mohammad was illiterate, it may have not been his fault, either, about what wound up in the final versions of his book.

But now, again, maybe you can see just how much importance Paul meant when he told Timothy to **"_Study_ to show yourself _approved by God_, a workman that needeth not to be ashamed, "RIGHTLY DIVIDING the WORD OF TRUTH"**!

And maybe now you can understand why "John" wrote the following words at the end of his book of *Revelation 22:18-19:* *[18] For I testify unto every man that heareth the words of the prophecy of this book, If any man shall add unto these things, God shall add unto him the plagues that are written in this book: [19] And if any man shall take away from the words of the book of this prophecy, God shall take away his part out of the book of life, and out of the holy city, and from the things which are written in this book.*

John knew first hand that even during his lifetime, '*contaminated*' words of misinformation inspired by Satan were being added to, or subtracted from, the words of TRUTH that were originally written by the 'inspired' authors of those books! Otherwise, he would have never written the last two verses he included into his book!

And Paul, too, knew that words he had not written were being added or subtracted to his books. That is why he implored Timothy, and thus, by proxy 'us', to really study his books, and by rightly dividing them, you would know the truth! So, my friends, it is so important that you study, not just each word in a verse, but that you do an in-depth study of all of the words, and books, and even, the authors of the books, so you are not deceived by Satan's inspired henchmen who were fluently proficient in inserting misinformation so as to lead you off from the truth!

One very important thing to remember right here; Satan is the father of all Liars, and the author of all Lies. Need proof? Just listen to the Lies that are spread daily from the mouths of the Lying politicians in our own government!

So, here, listen up! This is very important!

There are **_'NO' original copies of any of the original books written by any of the authors of the Bible in existence today._** Or at least, any that have been found at this time. The earliest known manuscripts of any of the books included in the Bible are some of the Dead Sea Scrolls that date back to between 1800 and 2000 years ago.

But as for New Testament books, the earliest known manuscripts, or portion thereof, is a business-card size portion of the Gospel of John that dates back to the early 2nd century AD. But the earliest of any of the copies of any other New Testament book dates back to the 3rd century AD. It was not until the Roman Emperor, Constantine, in the 3rd century AD, declared that Christianity was to be the authorized state religion, did any of the individual books of the Bible begin to be gathered together.

(If you are interested in more information about this, do an internet search for the 'First Council of Nicaea' that took place from May to August in 325 AD. It was named after the town where the Council took place, Nicaea, which is now called, Iznik, in the country of Turkey.)

So, therefore, it is impossible to know for certain that what you are reading in the "New Testament" *only* are the absolute original words originally written by the original authors.

But there is a solution for this.

**"But when he, the _Spirit of truth_ comes, he will guide you into _all the truth._" John 16:13**

Ask the Holy Spirit to guide you into 'all truth' as you earnestly study the words of Truth! He *will* answer your prayer.

Listen, if you're a Bible believing Believer, please don't take offence with, or think I'm bashing you for your beliefs that the Bible is the absolute word of God, inerrant and infallible in all of its aspects. It's not really your fault. You have basically been brain-washed into believing that the *English translated version* of the Bible as being inerrant and infallible because you've had to swear to that!

But just because the Elders of your church may have shiny sparkling degrees of higher education hanging prominently on their office walls don't mean they're always correct. They are 'human' teachers, preachers, and educators, after all. And especially, they are not the 'original authors' of any of the books of the Bible.

So, that is the exact reason why God inspired the Apostle Paul to include 2nd Timothy 2:15 in his writings. For you to be approved by God, you must be diligent enough to allow the Holy Spirit to guide you into all Truth, even if it varies from what the religious leaders you sit under say and teach! Be approved by God and don't worry about what the rest of the world, or your church, may say and do!

And what I consider a tremendously huge and satanically inspired subterfuge, is the requirement instituted by so many, if not all, denominations requiring you to publicly state your belief that every word in the entire Bible is inspired and inerrant from God, before you can be a member of their religious order! I truly believe this is Satan's direct attempt to keep you from performing your required duties of 2nd Timothy, chapter 2, verse 15 of studying the Bible in such a way as to be *"approved by God"!!!*

In other words, to draw the context of believing that every word in the Bible is inspired by God and inerrant in it's form, leads to the conclusion that there is no need to actually study anything written in the Bible beyond what each word actually states! Thus, if you fall for this satanic trickery, you will never actually be approved by God, and Satan can, and will, keep you scripturally ignorant and spiritually weak throughout your life, depriving you of so many blessings!

Okay, so now we all agree, or should, that

the timeline of Creation had its beginning approximately 13.799-billion years ago.

So, what happened after that?

# **CHAPTER 2**

## *"Light"*

Let's go ahead and answer one of those questions I presented in the Introduction section. And thank God Almighty, or the Cosmic Happenstance non-entity if you'd rather, that this is a question that both Science and the Bible have both agreed on and answered correctly.

The question is: ***"Was "light" really what created the Universe?"***

Yes! It really was!

So, how do I know that Light was what created the Universe?

Let me explain it this way by using the actual *facts* garnered from those dastardly scientists. (Listen, they're not all doe-eyed, puppet-master controlled, political peons or die-hard blinded Bible blasters. In fact, I would guesstimate that around 50% - 60% percent of the time, we can probably even believe them.) Such as in this case where they actually do, factually, tell us the truth!

The undeniable fact is that just a brief fraction of an instant before Creation happened, something totally impossible and completely unexplainable in any kind of a *natural* environment, a teensy, tiny, almost impossibly small miniscule spectrum of a form of "Light" somehow formed.

But, listen up here! This is very important because it will show you just how impossible in the *'natural'* that the actual beginning of Creation really was! Because, that actual spectrum of 'Light" was so tiny that it was actually smaller than the size ***of a millionth, of a millionth, of a millionth, of a millionth***, of a **single atom!**

***WOW!!!** That's r - e - a - l - l - y small!*

But what made this teensy tiny little speck of light to somehow appear in the first place is the completely indescribable, and even, unimaginable fact because, again, that light formed out of complete *"NOTHING"* in the *'natural'!*

*"HUH?"*, you gasp!

Yeah! *'Somehow'*, even though absolutely *"nothing"* in the *'natural'* existed at that time for it to form out of, somehow it still did.

I'll wait a second while you get your tongue back in your mouth and you put your socks back on.

Okay, now, I'm going to tell you something that is every bit, (or even more so, in my opinion) as amazing and *unbelievable* as the indescribable anomaly of a speck of light suddenly forming out of totally *nothing*. It is the indisputable truthful fact proven by science that a brief fraction of an instant after that little bitsy speck of light somehow impossibly formed, it suddenly exploded outwards *in all directions* in the biggest explosion to ever happen in the last 13.799-billion years since *history* burst into reality!

Listen, even if you gathered all of the nuclear weapons in the world and then suddenly exploded them all at the same time, that explosion might possibly be big enough to destroy the entire Earth, but in comparison to the explosion of the Big Bang, it would be nothing larger than what a droplet of water falling into the Sun would cause! In fact, if you could somehow capture the unimaginable destructive power of every Supernova that has ever exploded since the beginning of Creation and then rolled them all into one huge bang, even all of that awesome power would still be no comparison to the unlimited power of what happened at the very instant of the Big Bang exploding!

Okay, let's go on.

We now know that at the very beginning of Creation, an explosion of such an impossible magnitude of power and majesty took place that it is still to this very day stretching the boundaries of what used to be a big vast empty void of complete *nothingness.* But that leaves the forever enduring question of, if *'nothing'* actually existed at that time, then just *'where did the "Light" come from?'* How could 'anything' form when absolutely 'nothing' at all existed?

Well, to the riotous ridicule and raucous mocking laughter of the so-called educated scientific Elites and the Hollyweird know-it-all ignoramuses, the only person and place anyone has ever been bold enough to give an answer to this seemingly impossible question was Moses, the author of the Book of Genesis in the Bible. Moses boldly and unapologetically declared that "God" *spoke* that Light into existence.

Yeah! You want to talk about what's in the power of *'words'*, huh?

But, Moses' bold description of how that speck of light formed is way more than Science has ever stated. In fact, their utter and complete embarrassing silence on this question has been absolutely deafening in their non-responses! With their averted eyes and turned, red-flamed faces, along with their sickly-looking body-language of slumped shoulders and sunken chests, limply implying that it is a conundrum that is just unexplainable, and if you don't bring it up, they won't either. So, by the utter and complete embarrassed *silence* of the stiff-nosed and tongue-tied so-called educated scientific elites, without question, by default, *this round must go to the Bible-thumping believing Believers!*

But let's venture onward to see if there may be some kind of a possibility for the so-called educated scientific elites to make some kind of a comeback to even the score, shall we?

Okay, here's the bottom line ultimate, mother-of-all-bombs (MOAB) question: Was the "Light" that somehow impossibly formed out of complete *nothing,* that then was quickly followed by the explosion of the Big Bang, just some kind of an astronomical, none understandable, nor explainable, *'Cosmic Happenstance',* as so many people believe? Or, was it as many others, including Einstein, that believe(d) that some kind of a Supreme Being far above us mere mortal beings in knowledge and intelligence, design and create this all? A popular term that the so-called educated scientific elites use is, Intelligent Design.

Well, let's see if we can figure this out, okay?

Science has factually proven that when the vast explosion of light took place, it exploded outwards *in all directions!* (**Note:** Even though the Bible doesn't explicitly say the light exploded outwards in all directions, the Bible's whole premise is that it did. You see, the Bible is explicit in that God wants to be at the 'center' of our lives. In fact, He wants to be at the center of all Creation.

Don't believe me?

Well, the Bible states that one day, Jesus Christ will return to Earth and rule the Earth for one-thousand years with His throne being in Jerusalem. So, for your sake, take a world globe and put your finger on Jerusalem in Israel. Then, look around it. Your finger, and thus, Jerusalem, is located in the very center of the globe! Even on the

Earth, God created a special place for ruling the Earth from the very center of it. It only makes sense that His Heavenly throne is surrounded on all sides by His other creation, the Universe!

So, it's very simple to prove for yourself that Science is correct about the Light exploding outwards in all directions. All you have to do is take a look at what is in the Universe. Everywhere you look are big round orbs of stars, planets, moons, etcetera. Plus, look at the 'spiral' round galaxies that flood the cosmos, like our own Milky Way Galaxy, and the spiral of the **Andromeda** Galaxy, our closest galactic neighbor.

*(Just a little side mark here to put the fear of God, or the Cosmic Happenstance non-entity, in you: The Andromeda Galaxy is racing directly towards our own Milky Way Galaxy on a direct collision course at a speed of **670,000 miles per hour** (1,078,260 kph)*

*Yeah, my friend, just think, every 24-hour day that goes by, the Andromeda Galaxy gets more than **16-million miles** closer to colliding with our galaxy! So, in approximately another 4-billion years, the two galaxies are going to have a colossal collision, and then, its goodbye Earth! So, I feel so incredibly sorry for all of those enlightened Elites who are* *trying their hardest to create a way to become immortal human beings, or a combination of human and machine, like a certain large automotive manufacturing magnate plans to become with his so-called 'neural implant'!*

*But just knowing that it will take another 4-billion years for this to take place should give you a relative reference of just how powerful the explosion of the Big Bang really was! And it should also give you a direct reference of the enormous size of what used to be the vast empty void of 'nothingness' before the Light exploded into it!)*

Anyway, back to my subject!

All of these round, or spiral items, plus so many other examples of round things in the Universe, are there for a reason. In fact, let's bring this down to the *'round blue ball'* we live on. No matter where you're standing on our Earth, there is no true 'up' or 'down'. In other words, if someone was standing on the true North Pole and pointed 'up', while someone else stood on the true South Pole and pointed 'up', they would both be correct. So, think of the Universe as being a gigantic big round ball! And in the very center of this big round Universe is where the very first thing in Creation took place!

An explosion of LIGHT!

So, this question has to be asked; "If the Universe is round, and Heaven really does exist right smack dab in the very middle of it, then, is God really invisible? Or have we just not yet been able to invent telescopes powerful enough to actually peer into the very center of this huge beautiful monstrosity?

Well, maybe one day we will be able to do that. Then maybe we will be able to say, "Hello God! Nice to 'see' you!"

But, let's move on and finally answer the so-called unanswerable *MOAB* question just to see if Moses just may have been right. So where did the 'Light' that exploded into the Universe actually come from? Again, absolutely **nothing** existed at that *time* ('*time*' is used here in a silly, foolish, colloquially manner because not even 'time' existed at that *time,* ☺ either). So, how could something like a brilliant burst of Light suddenly form in the first place?

Okay, let me bring this all the way down to the complete lowest denominator. By doing that, it will give us the *'only possible answer'*! Okay, here it is; for 'anything' at all to be able to form out of completely 'nothing', then by absolute necessity, it can only be described as one thing; it had to be some kind of a _supernatural_ "MIRACLE"!

So there. That's it! There is no other logical answer than it being a *supernatural miracle.*

Listen, by definition; anything that takes place that is *'super'* – *'natural',* it has to mean that something *'super',* or far above the *normal,* had to have taken place far above the *'natural'!* So, the 'only Being' above us 'normal' beings, must again, by definition, be a *'super' Being,* or in more logical terms, a *GOD!*

But even so, there are a lot, *(HUGE),* number of people, some of them even somewhat intelligent, I suppose, that truly believe that this light suddenly formed and exploded into existence just because of some kind of a weird *'cosmic happenstance'!*

Well, it's okay to me if those people want to believe this. But I'll tell you one thing, I, personally, would never bet on the same horse that those people are betting on! To me, the odds of a so-called *cosmic happenstance* just *'supernaturally'* happening out of complete nothing, meaning that there was absolutely no way possible in any scenarios at all in the *'natural'* for it to happen, are just too astronomically great to ascertain a possible solution even with all of the *'super'*-computers' in the entire world hooked together and running the same unanswerable search!

Listen, all of you Cosmic Happenstance happening people, to deny the indisputable fact that something 'super'-'natural', or much higher than the normal, took place really makes you look absolutely silly! More bluntly, it makes you look like a silly fool!

But even so, I'm not going to try to be mean on purpose to you. But then again, in continuing to tell *The True Real Story of Creation*, I do have to take this *"Miracle of Light"* suddenly appearing out of complete nothingness even further, deeper, higher, far out into the **supernaturally miraculous** realm, again.

So not only did this teensy speck of Light, that was so incredibly tiny that it was actually teetering on the very slippery razor-sharp edge between finite and infinity, suddenly somehow appear out of absolute "nothing", but, it also, again somehow impossible beyond any explanation in the 'natural', is the unfathomable fact that inside of that almost impossibly small miniscule particle of "Light", was *"everything"* that *has ever, or ever will,* exist in the entire Universe!

That's right!

*Everything* that has ever existed in the entire universe was somehow bundled up and contained *"INSIDE"* of that teensy little barely visible flash of Light! So, when you look up into the sky at night and see the *uncountable* number of stars twinkling back at you, *everything* that made those stars was first contained inside of that teensy little speck of light that was almost impossibly small to even imagine!

Listen, I'll give you a small example of what kind of an incredible impossibility that was. Take a look up at the three straight-in-a-line stars that make up Orion's Belt. Now, directly above on the left side of that belt you will see a bright red glowing star even to the naked eye. That star is named, Betelgeuse.

Now here's just this one example that hopefully, will not hurt your mind so badly that you wind up in a round rubber padded room sucking your thumb and mumbling incoherently. But Betelgeuse is so large in size that if you could somehow switch it out for our sun, neither

Mercury, Venus, Earth, Mars, or the asteroid belt would even exist because, Betelgeuse is so large it would swallow them all up. In fact, its diameter is so large that it would almost extend all the way out to Jupiter! That's how much 'matter' is contained in just that one star alone! But yet, beyond all 'natural' possibilities, all of that matter in just that one star was once bundled up inside of that teensy tiny little speck of light that birthed Creation as we know and see it today!

So, is there anyone out there who wants to even attempt to give me another explanation of how that incalculable oddity of creation could possibly have happened other than it being another supernatural miracle?

I'm waiting . . . . Your silence is again, deafening!

Okay! But, let's take this supernatural miracle a step even further than Betelgeuse. Now, just think of all of the other stars, planets, moons, plus all of the uncountable astronomical number of other things that is out there in the Universe. Then, you must also realize that all of those other uncountable items were also once contained inside of that teensy little speck of light that was so small that only a quantum microscope would have been able to just get a whiff of illuminance from it.

But let's take this even another step further and bring it down to just our own human level. Take a look around at *everything* that concerns you. *Everything* that you see in the form of cars, trains, airplanes, houses, brother, sister, mother, dad, wife, dog(s), husband ~ *LIFE*, ~ *etcetera*, was somehow also impossibly originally *contained inside of that teensy little particle of Light.*

**But listen now, this is very important:**

Ever since the Big Bang took place well over 13-billion years ago, not even one speck of dust, or whiff of gas, (*thank goodness*) or a drop of water, or even another little sparkle of Light, has ever again even leaked, much less exploded, into the Universe!

So, let's all hear it one more time, *EVERYTHING* (all of the 'matter') now in existence throughout the entire Universe was once included inside of that first little speck of Light that suddenly, impossibly, came into existence from absolutely *"nothing"!!!*

*So again, not only did the spark of Light just appear out of __nothing__ at all, but all of the matter* *that has ever existed in the entire Universe, also just appeared out of **nothing** at all, and miraculously, all of it, too, was somehow contained inside of the teensy little twinkle of LIGHT!*

So, let me ask you another MOAB question. Just what kind of mathematical odds do you think it could possibly be for something of this *magnificence*, this *marvelousness*, and without a doubt, this totally and completely **miraculous,** to happen?

No one to my knowledge has ever tried to put odds on it, because undoubtedly, it would have to be an absolutely *IMPOSSIBLE* equation to determine! Mainly because no number mankind could ever come up with could contain that many zeros! In fact, I would venture an educated guess that if there were some such number, the zeros would, by necessity, stretch all the way across the entire still expanding matter-filled Universe!

So, let's do a quick review, shall we? That teensy little miniscule speck of 'Light' that was millions times millions times millions times millions of times *smaller than a single atom,* (so *small* that it would take an "Atomic Quantum Microscope" just to get a faint glow of it) suddenly appearing out of complete nothingness was the first *supernatural miracle!* Again, there was *"nothing"* there for it to even be able to form!!! There was *nothing* there to make the Light! But yet, it somehow impossibly still did! And it somehow formed out of complete *nothingness!* Which could be <u>nothing</u> else but a <u>*supernatural miracle!*</u>

Then, the second *miracle* was that teensy little cosmically small speck of light could by itself, suddenly burst forth in such an astronomically huge explosion of such majestic proportions *in all directions* that the light from it is still to this day, 13.799-billion years later, is still traveling at more than 186,000 miles *per second*, or close to *670,000,000-<u>million</u>-miles-an-hour!*

In other words, if you could somehow travel at this speed, you would be able to go all the way around the Earth 7 ½ times in *only one second!!!*

Now that is pretty fast, I would say! And I hope you skeptics would, too. **Note:** I won't get into it here, but if you want to see something that is really, really, really *FAST* that will surely blow your socks off, search the internet to find out just how fast that first speck of light was going at the very instant it exploded into existence at the Big

Bang!

**Hint:** The best guesstimates by man (Max Planck, founder of the Max Planck Institute) have still not been able to determine its exact speed; the main reason is because it's really another *supernatural* miracle because that explosion was traveling at the speed of ***"INSTANTLY"!!!***

Or more precisely, it was traveling at the speed of *God's words! God said;* ***"Let there be light! And there was light!!!")***

Or to put it in another 'light', God said, "Let there be light! And Light 'already' was!"

Oh, okay, stop your begging, will ya? I'll give you just one quick example of *approximately* what the 'speed' was at the very beginning of The Big Bang!

Now, you really have to use your imagination here, okay? So, imagine that before the explosion of The Big Bang, the spectacle of Light that included the entire complete Universe inside of it was the size of a golf ball, instead of it being an almost impossibly small speck of light. Now remember, that's everything in the entire Universe somehow now condensed and crammed into the size of a golf ball. But then, in the very same **'instant'** that The Big bang exploded, the entire complete Universe **had already** exploded outwards from the size of the golf ball to the size of the entire Earth!

In other words; the initial blast of the Big Bang was traveling outwards in all directions at the speed of **'instantaneously'!**

The Bible gives a much better example than the one above. In **1 Corinthians 15:52,** the Apostle Paul wrote; **"In a flash, in the twinkling of an eye, at the last trumpet. For the trumpet will sound, the dead will be raised imperishable, and we will be changed."** (New International Version; NIV) Biblica' Zondervan USA

Now that, my friends, is the speed of God's words! That is why it's speed is incalculable! It's just impossible to calculate the speed of anything that travels at the speed of *"instantly"!*

Okay, let me get back to my subject. Finally, the third, and without a doubt, the most outrageous and outlandish of all of the *supernatural miracles*, is the indisputable fact that teensy little cosmically small speck of light barely visible even with the most powerful microscopes, still had the capacity to contain **_EVERYTHING_** inside of itself that now surrounds us, plus another 7-billion people besides ourselves on just our little planet alone, plus including **_EVERYTHING_** else that makes up the uncountable number of stars, planets, asteroids, comets, and everything else that makes up the entire Universe!

So, my friends and the Cosmic Happenstance happening people, because of these three *supernatural miracles* alone, the mathematical odds for me to believe that just some kind of a weird, indescribable, and totally impossible *cosmic happenstance* suddenly somehow came together, and then following that, suddenly exploding into existence that could eventually, billions of years later, form "you and me", and everything else around us, are just too great for me to even entertain that being a possibility!

To me, only a foolish 'schlemiel' could believe this! There are just too many unexplainable things that can only be called "SUPERNATURAL MIRACLES" that happened at the very beginning of Creation for there to be anything other than **Someone** much greater and infinitely more intelligent than you and I to put this wonderment of Creation together!

So, what do I believe?

Well, I basically have no choice if I truly want to remain relatively sane, at least according to my wife's low standard of me, but to believe in a great and loving supernatural supreme "GOD" who so desperately wanted something so much more to admire and *love* than what existed in His own kingdom that was at that time surrounded by a completely dark and empty vast void of NOTHING! So, in His infinite wisdom and unlimited *supernatural* ability, He decided to create a place where indescribable majesty, beauty, and "LIFE" could happen, and He would be surrounded on all sides by it!

But, let's don't stop here. The story of Creation is so much larger than this. Remember, the explosion of "Light' and 'Life' into the Universe happened approximately 13.799-Billion years ago, or a really long 'yom' ago. So, we still have a lot of time to cover!

# CHAPTER THREE

## "Formation of the Early Earth"

The Bible says that the Earth was created on 'day' 3. But as we've already determined, 'day 3' should have been translated something like this; *"After a long extended period of time, (yom) the Earth was created by God."*

However, Science has determined to a pretty accurate degree just how long that extended period of time was. Here's their explanation.

Approximately one-half billion years after the Big Bang, and soon after the great slowdown in the speed of *instantly* that the 'Light' had been traveling at until it finally reached its constant unwavering speed of 186,282 miles per second, our own Milky Way Galaxy began to form, which by the way, is still evolving and expanding to this "day". But it took around another 9-billion years after the Milky Way Galaxy began to form that during a *'yom'*, a large amount of Helium and Hydrogen gases, along with a smaller mixture of a few other gases, all collated together and eventually formed into a huge round ball. Then, the huge gravitational weight of all of those gases pressing down against each other caused the inner core to heat up to somewhere around ten-million degrees **Fahrenheit.** Then suddenly, following that, an atomic reaction of nuclear fusion took place that caused the huge, gas-filled ball to suddenly burst into bright yellow and orange flames.

Following that, the huge flaming round ball began radiating not only a tremendous amount of heat, but it also began projecting a tremendous gravitational field around it for a long way out into the huge, rapidly filling no longer empty void. This gravitational field soon caused a tremendous amount of the rocky debris floating around in the vicinity of the big flaming ball to eventually begin to coalesce into four new round balls that were beginning to rotate around the big yellow ball in huge concentric circles.

Then beyond these four new round rocky balls, four other huge round balls made up mostly of different kinds of gases began to form, all of them in beautiful colors and magnificent designs of utter beauty! And following these four huge round gas-filled balls were some much smaller round balls made up of more rocks and different kinds of frozen ice.

And finally, there is a very great possibility of another huge round ball of some unknown composition that is currently being called, Planet 9. But, it has yet to actually be discovered by astronomers, although many computer simulations are saying that it does actually exist because of the gravitational effects it is having on some of the huge gas planets, and other floating detritus located out near the far edge of our solar system.

Plus, there is also a newly discovered mass of some kind of matter that is actually going against all known possibilities of physics. While all known planets in our Solar System rotate around the Sun in a relatively flat plane called the *solar plane,* this new mass is actually circling the sun at an angle almost perpendicular to our solar plane, plus, it is also circling the Sun in a backwards circle, too! Physics says this is impossible for this to be happening caused by the gravitational forces from the Sun!

But this is one of the reasons for the supposition of the existence of Planet 9. Astronomers say the only explanation that would give a possible explanation for this phenomenon is that something relatively huge that is also circling our sun but is located way out in the vast, no longer empty void, somewhere way past Pluto is causing this oddity. In other words, it would have to be a new planet that is being called at this time, Planet Nine. A planet that very probably takes approximately 15-thousand years for it to make just one circle around the Sun! It will be very exciting if and when our telescopes finally do pick it out of the huge number of background little points of light and prove that it really does exist.

But all of these round balls circling the Sun would eventually be called Planets. And these Planets, along with the big round flaming ball that would eventually be called the Sun that these planets rotate around would all eventually be called, the Milky Way Solar System.

Now, I know that some of you die-hard "Creationists" out there are still not fully convinced that just by changing one Hebrew word can fully answer the question of 'when' Creation actually began. I know that some of you are still convinced that only around 7- to 10–thousand years ago is when God created everything; not the 13-plus billion years I previously stated as being the actually age.

So, I want to give you the final definitive answer to settle all possible arguments.

Here goes; if you look this up on Wikipedia, you will get a huge amount of barely understandable scientific jargon speaking about several different ways the approximate age of the Universe has been determined. And believe me, it will give you a splitting headache if you try to read it all and understand it. In fact, it almost made me go to my 'safe place' and pick up my non-triggering coloring book to find relief and comfort from this big ole bad world of such exploding knowledge!

So, instead of inadvertently triggering any of you, let me just say this, hopefully, in a very easy to understand way. The age of the Universe is very easy to understand by using a very basic mathematical formula; **_Time_ = Speed of light over Distance.**

What this means is, first of all, we know that light now travels at a constant rate of speed, and of course, we know what that speed is. So, we can measure the distance that something is from our Earth by measuring the *time* it took for the light we are seeing now at this time to finally reach us after it left whatever astronomical source we're looking at. Thus, if we use the distance an object is from us, and the time it took for the light from that object to reach us, we can 'back date' the *age* of that distance, or in other words, *'time'*.

**Example 1:** Remember the star we mentioned earlier named Betelgeuse? Well, since we know the constant speed of light, along with a fairly accurate distance Betelgeuse is from us, we are able to fairly accurately say that the light we see coming from that star when we stand and look up at it actually left that star 640 'light years' ago. In other words, the light we see coming from Betelgeuse at this time in our history is actually 640 years old! That, my friends, is before the supposed discovery of North America by Columbus. In fact, it's probably around the time the Knights Templar were hiding treasure on Oak Island.

**Example 2:** Not too long ago, light was detected coming from an object that took over 13-billion years to reach us. Thus, we know just from that spectrum of light that the Universe is over 13-billion years old.

Now, I know that still, some of you are going to say; "Yeah, but didn't you say that when the Big Bang happened, light was traveling at the speed of 'instantly'? If that is so, then maybe that explains why Creation really only took place just 7- to 10-thousand years ago."

Good point, my friend.

But here's why that couldn't be. You see, when God created the Universe, He placed into it what we call, the *"Laws of Nature"* or, *"Natural Laws"*. So, we know that one of those *natural laws* is the *'law of Gravity'*. And we also know that the *law of gravity* is required for 'matter to appear. In other words, the *law of gravity* was required to pull the 'matter' out of the light so stars, planets, and all other things that make up the Universe could eventually form.

**Example 3:** Think of gravity as being a huge magnet. Now, if you place something in front of that magnet that the magnet can attract to itself ~ re: iron, for example ~ it will pull it towards itself. But, if you place the iron and the magnet just far enough away from each other so that the magnet can no longer pull that iron towards it then, it is useless.

In other words, think of the iron and the magnet both being inside of that initial blast of light that exploded out of the Big Bang, which in actuality it really was. So, as long as that light was exploding outwards at the speed of instantly, nothing could extrapolate itself out of it so the iron and the magnet remained traveling together inside of the Light at just the same distance that could keep the magnet from pulling the iron to it. The light had to slow down enough for gravity to materialize and begin to exert its influence on everything else inside of that Light.

Now, I know that you can still say; "Yeah, but God could just make everything in the Universe to just appear."

Yes, He could have.

But, He didn't!

Because, if He would have done that, then there would be no need for any *'natural laws'* to exist. And if that were so, we wouldn't be here. 'Natural laws' are what keeps our Earth traveling around the Sun at just the right distance from the Sun for 'life' to exist. It is also why we don't just fly off the Earth into space. Also, that would mean that for every star, planet, moon, human being, and/or, anything else that has ever existed in the history of the Universe, God would have had to 'speak whatever it was into existence' *each time*!!!

You get that?

Just think how busy He would be just creating each insect!

Listen, again, God created Creation only 'one-time'! And, as we've already seen, everything needed for Creation to become what it is now was already included inside of that Creation! So, here's the final answer to this: God has never created *anything else* 'ever' since He first created Creation! *"Everything"* in existence now, or has ever been, was already inside of that first, and only, Creation!

I will explain this further, and show just how God uses His *'natural laws'* He placed inside of Creation to create whatever He wants created now. An example is when God created the *'lesser light'* that He stated He was going to create. Or, the Moon, as we call it!

That is why God put *Natural Laws* into Creation so they could *reproduce after their own kind!* So just stop a minute and think about just how important *'natural laws'* are. They are what caused the very first particles of matter to separate from the Light, and are responsible for everything that exists in the Universe now, and will be what is responsible for every new particle

of matter to ever form into something else.

**Bottom line:** God created *'natural laws'!* And He created them so that the Universe would be more than just huge flashes of light exploding outwards in all directions. God created it so that it would be a *'self-sustaining'* Universe, with *'everything'* in it reproducing after its own kind!

I challenge you to do an in-depth study of astronomy and see just how stars are formed, and from those same stars when they eventually use up their nuclear fuel and their lifetime comes to an end, how the 'supernovas' they produce when they explode create so many other items like planets, moons, comets, etcetera.

Hopefully that settles that. So please, let me get back to my original subject about the creation of our Solar System.

As the planets in the Milky Way Solar System finally began to settle into their captured gravitationally determined orbits caused by the huge magnet we call the Sun, another unexplainable *supernatural miracle* would soon begin to take place. And it would be on one of the first four rocky planets.

That *supernatural miracle* would be called *"LIFE"*, and it would eventually somehow burst forth and cover the planet in so many different species of such beatific colors and designs that mankind has still not seen them all.

But which of the first four planets would it be?

Well, it was easy to rule out the first one closest to the 'Sun'. It would just be way too hot for anything to ever survive, much less evolve in the first place. But the next three all had a possibility. Especially the 2nd and 3rd planets. Those two began to form almost like identical twins; both of them initially having much of what would be needed for life to begin. It would be interesting to see what would be the deciding factor.

But as this was going on, something very sinister was happening that would have a totally devastating effect on the new Earth, and eventually, on mankind.

# CHAPTER FOUR

## "The War in Heaven"

*King James Version*

*Revelation Chapter 12; Verses 7 – 9*

*And there was war in heaven: Michael and his angels fought against the dragon; and the dragon fought and his angels, and prevailed not; neither was their place found any more in heaven. And the great dragon was cast out, that old serpent, called the Devil, and Satan, which deceiveth the whole world: he was cast out into the earth, and his angels were cast out with him.*

Friends, I truly pray that by now you rest firmly in the belief that "God" was the author and creator of Creation. I pray that your *logical mind* will tell you there were just too many unexplainable *supernatural miracles* that took place for there to be anything other than a God that performed the miracles! And if you don't believe in that fact, then I challenge you to prove to me, and to everyone else who reads this book, through the use of mathematics how else Creation could have happened? Listen, being totally frank with you, mathematics is the very thing that 'proved' to me that it had to be a Supreme God Being who could have created Creation.

But hang on, because there's even much more mathematics coming on just a little later that will leave no doubt at all that it could have been anything other than a 'God'!

But, at this point, I pray also that you have reached the conclusion that Science and the Bible are both to be believed, and that there are many mutual areas of reconciliation, especially if you *rightly divide the words of truth'* in the Bible. Because, you see, if you can believe that Science and the Bible are both correct in many areas, then it is just a small step for you to accept that Someone much greater than you and I made all of "this".

So, for much of the following narrative, I am going to bring much of the belief in a Supreme God into it such as the Scriptures written above. So, for lack of any better proof, I am going to give you the T. W. Manes explanation to describe The War in Heaven.

Please allow me to unabashedly insert here an excerpt from another book I previously wrote called, **"The Never Ending War".** I think it will provide a better and more understandable review of it because it brings real, easy to understand reasons for the war that took place in Heaven.

## *Chapter 4*

*A brief moment ago in the eternal timespan of the immortality of Heaven, I AM (God) sent a request throughout the entire city for every angel to stop what they were doing and to listen up. Once they settled down, with his voice flowing and carrying throughout the huge place, He then said to them, "My beautiful, magnificent angels, I have some wonderful news to give to you. But before I do that, I need to warn you that the news I AM about to give will, with certainty, multiply the amount of work you do for me many times over. But at the same time, it will also give each of you a greatly increased sense of purpose and direction, and most of all, it will fill your entire being with indescribable joy.*

*"First, my dear faithful workers, I AM no longer going to be just Ruler of our glorious home here in*

*Heaven, but I AM also going to be the 'Creator' of an entire Universe; a Universe that will include innumerable multitudes of different kinds of Light and Life! But here is the best part, my faithful servants; I AM going to make some of that Life in my very own image and likeness. I will call that particular Life, 'Man'. And 'Man' will be so special to me that I will invite them to become my very own children, and to those who will accept my offer to be my children, I will be their loving Father.*

*"Yes, my precious angels, I AM going to speak words and the eternal 'Light' that emanates out of my own being is going to blast forth out of my mouth in an enormous explosion of indescribable power and incredible majesty in all directions so that it will totally surround our home here in Heaven with brilliantly burning stones of magnificent Light.*

*"Even though this may be beyond your capabilities to understand my conceptions of quantum physics and mechanics, my angels, but this brilliant Light will incorporate every wavelength, color sequence, and the chemical and metallurgical physical characteristics in our entire Heavenly rainbow, so as it blasts forth, it will begin filling the entire huge, vast, empty, black void that now surrounds our home here, with my own eternal LIGHT.*

*"In other words, I AM going to fill the entire vast empty void with 'MYSELF'!!! This way, no creation of mine can ever say I couldn't be found or I didn't exist! I will be everywhere lighting their path to me if they will only open their eyes to see and their ears to hear!*

*"But this, my precious angels, is the most awesome and magnificent aspect of all! You see, to actually create the entire Universe and everything that will ever exist inside of it, all it will take is just a millionth, of a millionth, of a millionth, of a millionth, of a single atom of my own Light!*

*"Again, all it will take to create the entire 'Universe', and everything that will ever exist in it, will be just a tiny infinitesimal fraction of a single atom of my own Light! This is why the creation of 'everything' will be called, a supernatural miracle! And it will be such a supernatural miracle that no 'man' will ever be able to 'prove' that the huge 'bang' that explodes the Universe into existence came from a mere 'accidental occurrence' of unknown cosmic forces!*

*"My dear servants, included in this enormous explosion of my Light into the vast darkness will be what are called, natural laws. Now, one of these natural laws will be called, 'gravity'. And over the course of thousands of millenniums, 'gravity' will*

begin pulling on the natural light that is included in the creation of the Universe. This will eventually cause the natural light to slow way down from the initial, indescribable and **incalculable** speed of the Big Bang, until all forms and manner of light will finally begin to travel at one uniform rate of speed. And from that time on this uniform speed will never vary. It will never again speed up, and it will never again slow down, and this speed will last for all eternity. From that time on it will always be traveling at a constant speed of approximately 186,282 miles per 'second' as it continues reaching outward filling the huge vast void with awesome examples of different kinds of light!

"But because of this huge slowdown from the initial explosive speed that takes place at the time of the Big Bang, it will allow the natural law of gravity to pull all of the different wavelengths, colors, chemicals, elements, and forces that exist inside the natural light, out of it, so these different 'natural physical items' will eventually begin to congeal and form into millions times billions times trillions times quadrillions of huge, enormous, round orbs.

"This is because the 'Light' I AM going to blast out into the vast empty dark void is going to be filled with every color and concept and sequence of light that will be possible in the natural order of physics *and chemistry. My angels, as you already know from looking at the Heavenly rainbow of 'Light' that surrounds our home here in Heaven, there will be millions times billions times trillions times quadrillions of different colors and concepts of Light flowing into the vast empty dark void. And to form some of these different colors and concepts, the different wavelengths of light will have to contain different makeups of naturally occurring chemicals and elements that will allow this gravitational tug to bring them all together into these huge round orbs.*

*But, my precious angels, after gravity pulls these huge round balls together, one of the jobs I will need you to do is to go out to many of them and set them on fire so the flames from them will continue to fill the darkness of the vast void with Light for billions of creation years. But one thing you must know, this will be a continuing effort on your part throughout all eternity because the natural laws of nature and the unseen forces of quantum physics and mechanics I AM going to place into my light, will continually form new orbs. And those new orbs will need to be set on fire, too. So, in other words, my brave and glorious workers, the job I AM giving to you will last forever!*

*"But as these fiery orbs are set on fire, the blazing light and the enormous gravitational weight of them will then pull other particles and elements out of my all-enveloping Light until they, too, congeal into all kinds of different orbs that will then orbit in huge concentric circles around the awesome fiery balls of light. And over the course of many billions of years, gravity will cause these round orbs to move into positions and form what 'Man' will eventually call, galaxies and solar systems.*

*"Some of these round orbs will be filled with different types of gases and they will shine brilliantly in magnificent colors and awesome shades, while others will be filled with different types of metals and stones. Some of the orbs will be just solid rocks. While still others will have a molten core way down on the inside of them so the fire below can work its way up to the surface and bring forth with it all of the elements and carbons necessary to make Life spring forth. These orbs will be called planets while the flaming orbs will be called stars. And there will be so many of them strung throughout the Universe that they will be as uncountable as the grains of sand along seashores!*

*"I will also make wondrous other objects that will have icy white heads with bluish-white glowing tails stretching out for millions of miles behind them as*

*they move throughout the vast solar systems. I will also make rocky round stones appear that will circle the planets called moons.*

*"I AM going to decorate some of the planets with beautiful rings of varying colors that encompass many of the different shades of my heavenly rainbow. These rings will float around the planets in huge concentric circles, and they will fill the vast darkness with the radiant glow and beauty that now surrounds my throne. Other planets will have bands of clouds of many different colors that will float around them, while others will be frozen lands of scenic beauty and wonder. Some will have flames buried deep inside of them and they will erupt with brilliant shining flows of molten rocks that will cover the land.*

*"Again, there will be so many of these different kinds of orbs that it will be impossible for anyone but the Trinity to count the number of them.*

*"But then, my faithful workers, after approximately nine and one-half billion years after the Big Bang, I AM going to create a tiny little blue dot of a planet. It will be called Earth, and Earth will be so special to me that I will make it my footstool.*

*"The Earth will orbit as the third planet from a huge*

*burning star which will eventually, through my influence, be named, Sun. My reason for this is because it will be named in 'symbolism' after my own Son. As the 'natural' Sun will shine down and freely give it's radiant, life-giving light upon the Earth, it will also be a symbol of my Son shining his own 'life-giving' and 'life-restoring' light down upon all of the Life that will be upon the Earth. The Sun's light will wrap the Earth in warmth as a symbol of my Son wrapping his arms around that Life and giving his warm love to them. And as the Earth's Sun will cause photosynthesis to occur which will cause 'Life' to sprout forth in abundance all over the Earth, this will also be a symbol of my Son giving 'Eternal Life' to all who willingly will accept it. The great light that will shine from the natural 'Sun', along with the eternal Light that shines in and through my SON, will be a symbol of our own Light shining forth and lighting the way to our home here in Heaven.*

*"The Earth will be the most unique and beautiful of all of the planets throughout the entire Universe. It will have vast white clouds that will float serenely above the land as if an invisible hand will be holding them in place. At times, these clouds will gather up huge quantities of moisture and then release the water back down upon the Earth which will keep all*

*life forms thriving.*

*"The Earth will also have vast blue oceans of water that will cover two-thirds of the planet. The rest of the Earth will be covered in land, with some of the land flat and barren like a desert, while other sections of it will have huge snowcapped mountains that reach their spires and summits upwards as in worship towards us here in Heaven!*

*"Upon much of the vast stretches of land, I AM going to cause all different kinds of grasses and plants to grow. All of these will sprout seeds so they can continue to reproduce after their own kind in a never-ending cycle of Life. Trees of all different kinds will also grow upon the land; some that will keep their leaves and be green all year round, and others that will lose their leaves at the end of every growing season. The trees that will lose their leaves at the end of every growing season when the cold season of winter sets in will be a symbol to all Life upon the Earth that their earthly lives must eventually come to an end, too. But the evergreen trees that keep their leaves and greenness will be a great symbol that 'Man' can also have everlasting eternal life if he willingly chooses to worship us and dedicates his life to us like you angels do.*

*"For all of the different kinds of living creatures I am going to fill the land with, I will cause fruits and vegetables, and seeds and nuts, plus all different kinds of grasses and vines to sprout forth in great abundance, so they can all be used for food. But before I create all of the different species that will cover the land, I AM going to fill the seas, oceans, rivers, and lakes with fish and other kinds of living creatures of every kind that can live and thrive in the different kinds of waters. The waters will also have great mammals swimming in them, some that will grow to such huge size they will be called Leviathans. And all of these creations will fill the oceans, seas, lakes, and rivers in massive abundance.*

*"But then, after a few billions of Earth years go by, and as the Earth finally settles into an ideal climate and ecological formation, I AM going to create my ultimate creation upon the Earth. This, my beautiful darling angels, will be the greatest Creation I have, or ever will, create. This creation will be called, 'Man'. And just like you angels, I will create 'Man' to be immortal and have eternal life. But his immortal self will be his soul, not his physical body.*

*"You see, just as 'We', the Trinity, are 'Three in One', Man will also be created as three in one. He will consist of a physical body that allows him to do almost anything he sets his mind to do. We will give*

_him the ability to 'think' and 'reason' and 'choose'. His mind is the second part. And the last part will be his 'spirit' which will be the part that lives forever. In his 'spirit', we will put a longing for 'us'. Even if he has never heard of us, he will still long to know us. If he finds his way to us, which we will make sure the path to us will be lit brightly to show him the way, he will spend eternity here with us in Heaven. But, if he rejects the opportunity to look for us and doesn't try to find us, he will spend eternity with the one who will shortly become the evil 'Deceiver'._

_"But to separate my creation of Man from all other creations, and to show how special he will be to me, I AM going to create 'Man' in my own image and my own likeness. Thereby, every time I look upon 'Man', I will see myself in them, and it will cause me to love them beyond any measure or any sin they may ever commit against me."_

## Chapter 5

**B**ut for the first time ever, _there was a problem in Heaven. As the Ruler of the city was describing all of the glorious things He was planning to do, the most powerful and highest ranking angel in the entire hegemony of the angels, and who was also the most astoundingly beautiful, intelligent, and_ *wealthiest angel in all of Heaven, was suddenly filled with strange feelings that he had never felt before.*

*This huge, beautiful and powerful angel was standing right next to the Ruler's throne as the Ruler was declaring to all of the angels what he was planning to do. In fact, this great angel had been standing next to the throne ever since the first moment that the Ruler had created him. The Ruler had created him specifically to be the guardian of the most priceless object in all of creation, which was the Ruler's throne. It was the most valuable object in existence because whoever sat upon that throne, was the ruler of 'everything'! And this huge, beautiful angel had valiantly and faithfully, stood guard over the throne up until this moment. By standing in the very presence of I AM every day that went by in the timespan of eternity, this angel had become I AM's right-hand angel. As such, I AM confided in him all of His plans and wants and needs, along with what was needed to govern Heaven. By doing this, I AM gave the angel full authority over all of the other angels in Heaven.*

*I AM loved and adored this angel with his entire being! He even named him after Himself. I AM is a 'Light-Bearer'! So He named this great angel, Lucifer, which means 'Light-Bearer'. I AM had filled*

*Lucifer's entire being with 'Light', just like He, Himself, was filled with Light. Lucifer's nickname became, Angel of Light, because his very being shone with it! But I AM also called Lucifer the bright and shining Morning Star! By I AM calling him this, He was proclaiming in front of every other angel just how much He loved and honored this angel! By calling him the bright and shining Morning Star, I AM was declaring that Lucifer was the first thing He thought about every morning!*

*And Lucifer loved I AM every bit as much as I AM loved him! I AM was like a loving and benevolent 'Father' to him! In a way, I AM was his Father because I AM had created him. Lucifer knew I AM loved him; in fact, he felt like I AM loved him as much as I AM loved his own Son, and very probably, I AM did!*

*Later on, I AM would inspire a man named Paul of Tarsus to write in what would become a Holy Book that, "God is not a respecter of persons".*

*In other words, God loves everyone and everything the same. The bottom line was that God loved Lucifer with his entire heart and Lucifer loved God back with all of his! They both assumed it would be an eternity of love between the two of them that could never be broken - - - until now.*

## *Chapter 6*

*W*hen *I AM created this beautiful angel,* he created him to be the most perfect of all of the angels he would ever create. He had to be! He stood right next to the throne of God in God's own throne room in the very presence of God! He never left God's side!

*So, God created him to be the absolute model of perfection, perfect in beauty, and full of wisdom. The same as it was with the walls that surrounded the city; every precious stone adorned the angel's body and they were mounted in purest gold. And in addition to all of this, God created the angel to be the director and conductor of the Heavenly choir and Heavenly orchestra.*

*A Book that would one day be inspired by God for a man named, Ezekiel, to write, would describe the magnificence of the angel in this manner:* **"You were the model of perfection, full of wisdom and exquisite in beauty. Your clothing was adorned with every precious stone---red carnelian, pale-green peridot, white moonstone, blue-green beryl, onyx, green jasper, blue lapis lazuli, turquoise, and emerald — all beautifully crafted for you and set in the finest gold. They were**

**given to you on the day you were created. I
ordained and anointed you as the mighty
angelic guardian. You had access to the holy
mountain of God and walked among the stones
of fire.**" *Ezekiel 28, New Living Translation Bible,
Tyndale House Publishing*

*Lucifer had to be adorned this way because God had
placed his own 'well-being' in the hands of this
beautiful, powerful, and wise angel, and God trusted
him fully and completely! And God's throne room is
so astoundingly brilliant and filled with such
indescribable wealth and beauty that everything
that was located in it had to be of enormous value
and beauty, too. Since Lucifer stood directly next to
the actual throne of God, God filled his being with
indescribable brilliance and wealth!*

*But now, as this brilliant and powerful angel stood
within an arm's length of his surrogate Father
listening to the words being spoken by Him, this
most trusted and loved of all angels suddenly felt
like he was being wrapped up and totally enclosed
in an ever-tightening cocoon of strange feelings and
smothering emotions that he had never felt before.
His astoundingly handsome face was covered in a
dark mask of confusion and intense anxiety. His
large powerful hands were trembling and sweating
so much that he could hardly keep the scepter of*

*God from falling out of them. His powerful and muscular legs felt like they were made out of soft rubber that was quickly losing the ability to hold him up. And although he had never before in his entire existence been in a situation so uncomfortable to actually make him perspire, at that moment, twin raging rivers of damp, sticky moisture were running sprints up and down his spine. A very small part of him felt a strange sensation he labeled guilt. Another larger part of him felt an even stranger and much more powerful sensation he called anger! But the biggest part of him felt a sensation he named jealousy!*

*Then suddenly, like a raging river of mass destruction, these three sensational feelings swept through, over, and around him until his entire being from the top of his head all the way down to the bottom of his soles were enclosed in a smothering cocoon he called, pride! And within seconds of that, his entire being was filled with EVIL!!!*

*Not long after that, these 'evil' feelings began making him feel something even more sinister. They began making him feel empowered and strong, as if he had the ability to stand up in the actual face of the Most High as an 'equal' to Him, instead of being subordinate to Him. And as these feelings were sweeping through and consuming him, his entire being was rapidly turning into a bursting beacon of absolute pure evilness, and he liked it! The feelings felt good! "Oh, so good!" He thought!*

*Then suddenly, like a bolt of lightning shooting through him, he felt the overwhelming enormously exhilarating feeling of what absolute power feels like. It made him feel like he was not only the same as God, but that he had risen above the True God to become GOD in place of Him! It was an intoxicating feeling! Never before in all of eternity had he felt as drunk as he was feeling right then. Never before in all eternity had he ever felt as strong and powerful as he did right then! And all of it was making him feel like he could steal, kill, and destroy everything this 'arrogant' God was going to create, if he so desired!*

*He had 'power'! Absolute power!*

*Never before in his existence had he been jealous! But this jealousy had completely turned the love he had for God that had so filled his entire being to overflowing, to becoming absolute complete hatred for Him now! And never had he felt guilty before. Although as the feelings of anger and jealousy and overwhelming enormous power and pride grew stronger and stronger until they were filling his entire being, the weak feeling of guilt quickly* diminished until it was eventually repulsed completely out of his system by the other, stronger, and more joyous feelings.

Within a short few minutes, he didn't care at all about God. In fact, where there had been only an overriding all-consuming abiding love for God that had no words to describe the depth of it, he now felt the same depth of hatred for Him! And he was relishing in the exhilarating sensations of these feelings that made him feel strong, invincible, an equal to God.

It was enough to make his head swim; to make his body feel all tingly; to make him feel like he should be sitting on the throne of the Most High instead of standing next to it, guarding it as someone way below his stature, wealth, and authority! He began thinking of a way to steal the Ruler's Throne! 'Maybe that could be possible', he thought. 'But if I can't do that, I will create my own throne and I will then place it above the throne of the Most High!'

As these 'wonderful' thoughts were racing through his now corrupted mind, for the first time in all of eternity, he purposely looked down at the purest, transparent gold of the street he was standing on. He wanted to see his own reflection in it. The gold was so pure and flawless that it reflected his *appearance back up to him as if he was standing in front of a full-length mirror. And what he saw filled him with even more 'pride'! He stood there staring at himself, basking in the feelings of such awesome power and the perfect beauty that was staring back at him!*

*And especially, he was basking in full-blown wickedness! These feelings had enveloped him and filled his entire being with the narcissistic elitist feeling of pure EVIL! And this evilness made him feel utter disgust "for this pompous, egotistical, self-righteous God who was so vain that He wanted to create another living specie in his own image and likeness!*

*"What audacity! What boldness!" He proclaimed to himself! "What an ego this Ruler has to want to make something in his own image and likeness. Whoever heard of such a thing? Who else but this repugnant God could ever dream up something about making another creature in his own image? What audacity!!!*

*"But what really sets me off and makes me hate Him so much now is that He wants to create something more beautiful and glorious than me! I mean, just 'look' at me! Up until now, I have been the most beautiful and wonderful creation He has ever made.*

*And I will not stand by and allow this pompous, narcissistic, egoistical God to treat me this way!*

*"But wait," he declared. "Maybe I won't have to stand up against God alone. I bet some of the other angels will go along with me; especially when I give them the reasons for it, and if I promise to give them their own thrones to sit on next to me after I put my throne above the Most High. If I can get enough of them to go along with me, we can overthrow this pompous God who has controlled and used us as his own personal slaves all throughout this past eternity up until now. And after I throw Him out of Heaven, I'll create a being in MY own image and likeness!*

*"Yes! Yes! I like it!" He cried!*

**W**ell, if I do say so myself, this is a pretty good *dramatized* description of why there was war in Heaven, and how eventually, the war wound its way down here to the Earth since Satan was thrown down here.

You see, the problem was that Lucifer, who quickly acquired the name of Satan, which means, *"The Deceiver"*, became insanely jealous because God planned on creating a living being in His own image and likeness, and Satan was afraid that God would love His Creation of Mankind more than God loved him. But as the Scriptures above declared, Satan quickly lost that first battle and he was quickly disposed of. He, and the angels he had *deceived* that had chosen to follow him in his rebellion, were thrown out of Heaven and **"were cast down to the Earth."**

But ever since that time, Satan has done everything in his power that he could possibly think of in his war against God, and thus against 'us' as God's proxy, to either destroy the Earth entirely, or to cause as much death and destruction upon it as he could! In the gospel of **John, chapter 10 and verse 10**, it says; **"The thief (Satan) comes to <u>steal, kill, and destroy!</u>"**

Because of Satan's insane jealously of God, and of his unfathomable intense hatred-driven desire to steal, kill, and destroy all of the 'replicas' of God here on the Earth, it means that 'we' have been placed smack dab in the very middle of this never-ending war! It means that all of us, every person on this Earth, now carries a huge target painted on our persons! And it unalterably means that 'we' have become the ultimate *'take home prize'* for either Satan, or God, in this never-ending war!

But thankfully, the victory dance over us is ultimately in our own hands. Neither Satan, nor God, has the final verdict over who wins and who loses! When God created mankind, He gave us *'freewill'*. Thus, *it's our own choice* of who will ultimately *'take us home'* with them!

The choice we have to make in our lives is whether 'we' will humbly bow our knee in Holy reverence of God, and will willingly accept His Son, Jesus Christ, into our hearts and lives as our Savior and Lord over our lives. If 'we' will do that, we will then have eternal life, and as our reward, we will get to spend that eternal life living with God in Heaven.

But if we willingly reject the saving grace of Jesus and live our lives unashamedly as an entrenched co-conspirator in the lying, deceiving webs of Satan, we will unfortunately become the *take home prize* of Satan where, as our reward, we will spend all eternity in Hell with him and all of his demonic followers!

We should all be thankful to God that He gave freewill to us because Satan has never, and never will, give up his attempt to destroy us, or destroy the earth! He never will until one day, God will put a permanent end to him and his foolish war!

# CHAPTER FIVE

## "Satan's Attempt to Destroy the Earth and the Formation of the Moon"

Satan was floating in space near the Sun watching the planets of the early solar system form. His attention was particularly focused on the 2nd and 3rd planets. They were so much alike; very similar in size, and both with much of the needed properties that Life would need to eventually evolve upon them. Many years later astronomers would say that in both planets' early infancy, they could have been called 'fraternal twins.'

At this time as the planets in our solar system were just forming, I don't believe Satan knew exactly on what planet God would actually place His creation of 'man'. I believe that all he knew was the name that God had given to the planet, Earth, but not its actual order in the solar system where the Earth would reside. So, I believe that Satan was keeping an eye on the most likely candidates, the 2nd and 3rd planets.

But all of a sudden, out of the corner of Satan's eye, he caught a glimpse of something streaking really fast and it looked like it was coming straight towards the second of these two planets! From way out in an area of space of what would eventually become known as the Oort Cloud, which is an area way out around the Dwarf planets of Pluto and Eris, a huge, irregularly formed partially round orb hundreds of miles wide was racing into the solar system, and it's trajectory was on a direct collision course with the 2nd planet.

Satan was utterly transfixed as he watched with wonder and amazement as to what was surely going to be a gigantic collision of epic proportions! With his eyes growing wide with excitement he continued watching as the huge orb finally blasted directly into the 2nd planet! . . . And as it did a huge smile crept across his face as an evil idea suddenly formed in his mind!

The drastic collision between the two solar objects was so massive and destructive that in an instant, the smaller orb was totally destroyed! And the 2nd planet came exceedingly close to meeting its untimely end, too. The extreme speed, the drastic force, and the kinetic energy of the devastating collision completely stopped the normal counter-clockwise axial rotation of the planet dead in its track.

Then, a short time later, a weird phenomenon began to take place that only one other planet, Uranus, in the solar system does. The planet actually began to spin backwards in a retrograde manner, or what is a clockwise spin. Unfortunately, though, the backwards spin of the planet turned into a very, very slow spin. Ultimately, it would settle into such a slow rotation rate that the 3rd planet would actually complete 243 twenty-four-hour revolutions for the 2nd planet to complete just one single rotation around its axis.

And in addition to all of this, the drastic collision knocked the 2nd planet's axial tilt over to a severe 177-degree angle.

The devastating effect of such a slow rotation and its drastic axial angle meant that the great furnace of molten iron located in the center core of the planet, just like the 3rd planet has, was no longer capable of spinning. The unfortunate result of this was that the molten core was no longer able to project a magnetic shield out into space around the planet like the 3rd planet has. The result of this was that without a magnetic shield it would no longer be protected from the Sun's scorching fiery heat, or it's blasting solar wind that travels approximately 1,864,113-million-miles-per-hour, and along with the deadly solar radiation that comes along with all of it.

It didn't take very long before all of the water that had been present on the surface of the planet, which was a necessary ingredient for life to evolve upon it, was quickly blasted off out into space, either by the devastating solar wind, or it quickly evaporated out into space from the almost 900-degree surface temperature that suddenly arose upon the planet because it no longer had any protection from the Sun's solar rays. Thus, the final result from the great collision was that the 2nd planet quickly turned into a very vivid virtual reality of what a molten fiery Hell very probably looks like!

As Satan watched all of this, his evil, demented mind began working overtime. He turned to his demons and said, "That collision almost totally destroyed the 2nd planet. In fact, it was so massive that it totally ended forever the chance for any life to ever arise upon it.

"So, listen, I'm going to do the same thing with the 3rd planet, which has to be the Earth. But, instead of the planetary object that crashed into the 2nd planet being only small enough to just damage the planet, I'm going to direct a much larger orb to blast into the Earth than the one that hit the 2nd planet, and without a doubt, it will destroy it completely!

"So, listen to what this will accomplish, my demonic followers: if the Earth no longer exists, then God will no longer have a place to put his so-called 'Man' upon it! Then, I will lead all of you brave warriors back to Heaven and we will kick that old worthless foolish God out of it. That old fool once made the statement; *"Out of thine own mouth will I judge thee, thou wicked servant." Luke 19:22*

"Well, it's simple to deduce that if God is unable to complete the creation of the Earth like He promised **with His own mouth,** that makes Him a huge liar! He decreed that *"all liars shall have their part in the lake which burneth with fire and brimstone"; Revelation 21:8!* And because He made such a foolish statement, He will bring His own sure judgement down on Himself!

"Then, once we throw Him and everyone else still there in Heaven out of it, I will assume the throne of God and rule all of Creation! I will then recreate all of Creation in my own image and likeness!"

He turned towards his warrior angels and ordered them to go out and search the cosmos for something he could use to destroy the Earth with. It was not long afterwards and they returned. But, in their haste to please their demanding and unreasonable master, the best they could come up with was a small rogue planet very similar in size to the 4th planet in the solar system. It was much larger than what had hit the 2nd planet so they thought it would be large enough to get the job done that their master desired!

But Satan saw what they were bringing his way, and he screamed! "Is that the best you could do? Look at it! It's no larger than the 4th planet!"

He was so angry he began slobbering all over himself! Eventually though, he gulped another lungful of air, and continued screaming; "Are all of you so stupid that you didn't know what I wanted? I wanted you to get an orb around the size of the 5th planet! Something really, really huge!

"Now, I don't know if this one is even big enough to cause any damage at all to the Earth, much less destroy it!" He whined!

He was fuming! Over the span of the last 9-billion years since he had first deceived these former heavenly angels into following him in his rebellion, and while they all were waiting for God to finally begin creating the Earth, he had totally begun to despise them. As far as he was concerned, they had all turned into a bunch of lazy do-nothings unless he rode their worthless rosy-red reptilian rear-ends hard!

He finally spat, "Well, since you worthless lazy moochers couldn't find anything any better than that one, I guess I'll have to use it."

None of them dared to try to offer an explanation or an excuse, and thus, personally face the Snake's renowned temper, which had been getting massively worse as these last *9-billion* years had slowly passed by. So, they kept their mouths closed and quietly followed him to where the orb was located currently running amuck in the asteroid belt located between the 4th and 5th planets.

But shortly thereafter, they watched in awe and amazement as the mighty, demented, demonic dragon used his great power to alter the rogue planet's speed and direction until he had put it on a direct collision course with the Earth; or at least of what he believed he had done. In actuality, he was off by just *"this much"*, which meant that the huge orb was going to hit the Earth at a glancing blow, instead of directly head on!

Even so, in a short period of time, Satan, and all of his demonic followers, watched in utter amazement as the rogue planet smashed into the Earth! And even though it was just a glancing hit, it was still a colossal collision!

And to Satan's very happy belief, from the initial looks of the devastating collision, it certainly appeared that he had just accomplished his goal and destroyed the Earth! And to his demented mind, it meant that he had just won the war between him and God before it really even got started.

But Satan had completely forgotten one very important item that God had said He was going to create along with the Earth. God had said that He was going to create a huge light in the sky to light the day, along with *'a lesser light'* to light the way at night. Well, the Sun was the huge light, but at that time in the infant Earth's creation phase, it was still without the lesser light to shine upon it ~ *that is until Satan's attempt to destroy the Earth.*

As God looked down upon the Earth at Satan's humorous attempt to destroy it, He smiled and thought, ***"What Satan meant for evil, I will instead turn to good."***

So instead of the rogue planet destroying the Earth as Satan had surmised it was going to do, in a short period of time of only approximately one-hundred years, God caused the massive debris field surrounding the spinning Earth to quickly coalesce into a large round off-white orb. And this new orb soon settled into the place where it began to orbit the Earth as it's captured moon!

But at the same time, God made sure that this new orb had a highly reflective surface upon it so as the light from the Sun hit it, instead of the new moon absorbing the light, it was instead reflected downward onto the Earth! Now, the Earth had the lesser light that God had promised He was going to create. So instead of God being turned into a liar by Satan, God used Satan's evil attempt to destroy the Earth and instead, turned it into 'good'!

**Just a quick reminder here:** Even though God had stated that He was going to 'create' both a great light and a lesser light, the way He actually created these was by only using what He had originally placed inside of that very first speck of Light at the beginning of Creation. In other words, he used the *'natural laws of nature' to* create both lights, and the entire Universe.

And to prove without any shadow of any doubt that it was "He" that had done this great thing, God again created a few major, totally unexplainable, *supernatural* miracles to prove it!

The first *supernatural miracle* God did was He caused the rotation of the Moon to settle into a direct *synchronous* rotation with the Earth. The effect of this is that the moon began to rotate around on its axis at the approximate speed of 10 miles per hour. In comparison, the Earth rotates around at an approximate continuous speed of 1038 miles per hour, or just over 17 miles *a minute.* So, in effect, the moon makes only one full rotation on it's own axis as it takes for it to make one full revolution around the Earth; or a little more than every 27-days. Thus, taking into account the different sizes of the two cosmic bodies and the distance that separates them, both cosmic bodies rotate together in such a way that the side of the moon facing the Earth is *always* facing the Earth. It never varies!

For something like this to happen in all of the vastness and randomness of space, the calculable mathematical odds of this 'supernatural miracle' have to be massively huge! (The uniqueness and 'orderliness' of this miracle is just one of the reasons why Einstein knowingly stated that 'Something', or 'Someone', far above us mere mortals had to 'Intelligently Design and create our Universe! This, along with all of the other unexplainable 'supernatural miracles' are mathematically impossible for them to happen by random!)

Although this *supernatural miracle* in the creation of the Moon was astronomically huge, another *supernatural miracle* is indisputably the greatest one of all; possibly the second greatest only in miraculous to the initial Big Bang!

Because of the size of the Moon, and the distance that the Moon is to the Earth in relation to the Sun, the Full Moon and the Sun appear to be the *exact same size in the sky*! The mathematical formula for this is that the Sun's diameter is 400 times larger than the Moon's, but the Sun is also

400 times further away from the Earth than the Moon is. So, this *'supernatural miracle'* of '400' and '400', is way, way, way beyond the calculable mathematical odds of human determination!

In fact, the mathematical odds of this *'supernatural miracle'* has to be equal to the mathematical odds of Light forming out of nothing, along with that teensy speck of Light somehow inexplicably being able to contain all of the matter that has ever existed in the entire Universe, in addition to that little almost infinitely small speck of Light somehow causing such a massive Big Bang explosion that the effect of it has caused the Universe to continue expanding outwards in all directions at the rate of over 6-million miles a minute for over 13.5-billion years!

But the final *'supernatural miracle'* of this, 400 and 400 mathematical ratio is that at certain times, the Moon will pass directly between the Earth and the Sun. When this happens, it results in a *total* eclipse of the Sun. The only things visible of the Sun are what look like sparkling little diamonds peeking out around the edges. These are called Baily's Beads, so named after Francis Baily who explained this phenomenon in 1836. (Wikipedia)

Much later on, God's Son, Jesus, would declare that *'signs and wonders'* would appear in the heavens before His return. A total eclipse is one of those wonders because Jesus declared that the Sun will be turned to darkness. Plus, another 'wonder' Jesus declared is that the Moon would turn to blood. This actually happens when the opposite of a solar eclipse takes place when the Earth moves in front of the Moon blocking the Sun's light. This is called a lunar eclipse which causes the color of the Moon to turn a deep, blood-red color.

And let me just add another 'supernatural miracle' to the moon and Earth's relationship. After the rogue planet slammed into the Earth, which caused massive destruction on the Earth and total annihilation of the rogue planet, an astronomically huge debris field was left over directly in the path of the Earth's orbit around the sun. Now, in what we consider normal astronomical circumstances, for the moon to coalesce into its current form and size out of this debris field should have taken hundreds of thousands of years to complete.

But, in actuality, the moon formed from this debris in around a total of only 100 years, which is less than an eye blink in astronomical timeframes!

So, to the limited finite human mind, these *'supernatural miracles'* surrounding the Moon are completely unexplainable and have absolutely no answer to define them in the natural realm. Therefore, the only explanation for any of them by themselves, much less all of them combined, is the uncontestable 'fact' that something Supernatural caused them.

I choose to believe that the Supernatural entity that caused these is a Supreme Being called Jehovah God! And I pray that you do, too!

So, what will happen next?

# CHAPTER SIX

## "Life Appears on Earth"

**A**fter God created the Moon out of Satan's debris field, the Moon began to do exactly what God had intended for it to do when He had said He was going to create it. Now, don't take this statement out of context and say that since it was Satan's debris field, that Satan was actually the one who created the moon. This is false in so many ways! Yes, it's true that Satan created the debris field. But, it's also true that God was the One who caused the debris to form into the size and shape of the moon, and He did it in an astronomically short time frame.

Listen, as I stated before, God took what Satan meant for harm and turned it into good. God believes so strongly in this form of creating supernatural miracles that He turns evil into good all the time. All throughout the Bible are stories of how God did these actual miracles in people's lives and circumstances! And if you were to do a completely honest search of your own life, I'm sure you would find over and over again, areas and solutions where God did the same in your life by turning evil into good! Bottom line, folks, God is GOOD! Satan is EVIL!

But let me get back to what I was saying about the moon. After God formed the moon and set its speed of axis rotation along with its orbital plane around the Earth, it began to regulate the tidal effects of the oceans, which in turn began regulating the climate on Earth. The effect of this was that over a long *'yom'*, the Earth entered into an ideal climate where 'life' could *naturally arise* upon it.

**An important fact:** Since God had first blasted 'Light' out of His mouth in the form of His words, *"Let there be LIGHT and there WAS Light"*, along with the undeniable fact that that 'Light' had contained *everything* that has ever existed in the Universe, it was not needed for God to come down to the Earth and create the fishes and mammals that rapidly began filling all of the bodies of water upon the Earth, or anything else that was evolving upon the Earth.

In other words, in the *natural laws of nature* that God had originally included within the Light of the Big Bang, God's total work of creating 'life' had already been completed at the very instant 'light' blasted out of His mouth. God didn't need to do or say anything else. So as the Earth evolved to a point where it could support life, life *'naturally'* evolved upon all of it.

The same thing could be said about when God declared; "Let there be LIGHT, and there was LIGHT!" Well, it would have been just as easy for God to have said; "Let there be LIFE, and there was LIFE!" He didn't have to create LIFE *'again'* because, He had already created it!

*Listen,* I know what I just wrote probably made a bunch of Bible-thumping born-again believing Believers furiously angry as you are yelling that this smacks directly as a form of Deism! But before you throw this book into the fireplace and burn the possibly dastardly heretical tome to ashes, let me ask you this; If God had not already placed the ability for life to 'naturally' evolve on its own volition upon the Earth, wouldn't that mean that God would have had to come down to the Earth each time a new specie, a new plant, a new tree, a new insect, etcetera, appeared?

Well of course it would! So please, continue to read a little further and it should make a lot more sense to you. So, in this aspect only, Darwin's theory of naturally occurring life was mostly correct. But even more importantly, when the Apostle John wrote his gospel named after him, he wrote in Chapter 1, verse 3 the following words: ***All things were made through Him, and without Him nothing was made that was made.*** Thus, John, too, was absolutely correct in his analysis! But you have to go all the way back to the *'natural laws of nature'* that God had already added to the mix inside of the Big Bang; not to the creation of each form of life on Earth!

**An example to prove this:** In Genesis chapter 2, Moses wrote that God came down to the Earth and created a man He called Adam. (This is the one and *'only'* time its recorded in the Bible that God came down to the Earth and created 'anything'!) But the point of this fact is that God **did not** speak Adam into existence like He had done at the very beginning of Creation. No! What He did was He created Adam *out of the dirt of the ground.* In other words, God created Adam **_out of what He had already created!_** He created him out of *nature!*

Actually, though, to be absolutely factually correct, God didn't actually 'create' Adam. The Bible says that ***God _'formed'_ Adam from the dust***

**of the ground.** ~ (I'll expand on this in much more detail a little later on, and also answer the questions of where Neanderthals and Cro-Magnum, and all of the other so-called members of the human evolutionary track came from.)

But during this period of life's early beginnings, the climate of the Earth was very different from what it is now. At that time, the amount and purity of the oxygen in the oceans, streams, rivers, and lakes, plus the atmosphere above, was so great that huge living species began to populate the Earth. One of the huge specie that evolved in the ocean would one day be called, Leviathan, in the Bible. It would be the 'fish' that swallowed Jonah.

Now, let's don't forget about Satan, and his declared war against God. One day, Satan happened to take a look around, and he noticed all of the huge fish and extremely large mammals swimming in the different waters. Once again, another evil idea suddenly began to form; *"Since I failed in destroying the Earth by crashing a planet into it, I'll instead concentrate my efforts on destroying that puny, worthless 'Man' that God is going to one day create in His own image and likeness. I'll just steal, kill, and destroy them all once He creates them! And since God was so extremely foolish by already placing all of the necessary ingredients into the Earth that I will need to accomplish this, I'm going to cause huge, gigantic reptiles to arise upon the Earth that will attack, kill, and then eat God's Man!*

*"And along the way, I will also cause some other species that will resemble my own persona in image and likeness. These will crawl upon the ground, and many of them will have large fangs in their mouths that will be full of poison like the poison I have in my own mouth. These also, will attack and kill God's worthless Man. In addition, I'll also cause huge flying beasts to evolve and cover the Earth that will swoop down out of the sky and attack and kill God's Man!*

*"Oh, this is going to be so much fun to watch!!!"* He screamed!

So, this is what Satan did. He became the author and creator of the Dinosaurs, the huge birds of prey, and the poisonous snakes made in his own personified image and likeness that soon covered the entire earth!

But to show just how limited his power, wisdom, and abilities really are, one of the largest and most fearsome looking reptiles to ever walk upon the Earth, the **Sauropods**, evolved to be plant eaters instead of meat eaters.

Satan saw this and screamed, *"Well, I'll be damned!"* (Out of your own mouth comes judgement! So please, don't ever, ever, ever use this phrase!!!) *"The biggest, baddest, freaking Dinosaur I've ever created eats plants! Un-freaking-believable!!!"*

But God was sitting back just smiling and thoroughly enjoying watching Satan's foolish efforts in this 'war' that Satan had initiated! He said to Himself, *"You old worthless Snake, you just wait until I, again, turn what you meant for evil, into good!"*

# CHAPTER SEVEN

## "The Dinosaurs Demise and Satan's *'Imitations'* Arise"

Every day that went by Satan grew more and more frustrated that God had still not yet created His Man in His own image and likeness upon the Earth! He couldn't **steal, kill, and destroy** what wasn't there yet! His fiery temper was just about at the breaking point! He screamed at his demonic horde; *"Demons, listen up! It's now been well over 4-billion years since the Earth was formed. And in addition to that, my creations of the huge reptiles and the poisonous snakes, and all of the other man-killing monsters that now roam and infest the entire Earth has been going on for around the last 180-million years now!*

*"Do You hear me?" He screamed! "I said, one-hundred and eighty MILLION YEARS! And yet, the Most High God has not even come down to the Earth to take a look around to even try to find a place to create His Man!"*

He lamented! *"Oh, what in hell's blazes can I do to entice Him to go ahead and create His Man so my creations of pure evilness can wipe them all off the face of the Earth?"*

For the entire 180-million years that Satan's creations of death had roamed the Earth, God had barely been able to contain Himself from the joy of sticking Satan's evilness right back down Satan's poison-filled throat. But He had created the Earth with several *Laws*; one of which was the law of *Nature*. And for as long as the Earth exists God is going to allow the Law of Nature to serve the purpose He had created it for! So, God knew that if He was really going to turn Satan's evilness into a great force for good, He had to wait for this extraordinarily long number of years to pass slowly by for the huge beasts to propagate exponentially and cover the entire Earth.

But now, the time had finally come. God exclaimed to Himself; *"You just watch, you old fiery fool! Everything you have done to bring havoc and cause death throughout the Earth for the last 180-million years is now going to come crashing down right on top of you!"*

God looked down into the asteroid belt between Mars and Jupiter and spotted an asteroid around six miles in diameter. Then, with a whisper of His breath, He blew gently on the asteroid and redirected the orbit of it, which put it on a direct collision course with the Earth. Shortly thereafter, the huge rock crashed head-on into the part of the ocean that would eventually be called the Gulf of Mexico, and the huge crater it created would be named the Chicxulub crater, after a small town now near its epicenter.

The impact of the collision was so massive that a worldwide catastrophe immediately began enveloping the Earth. Within a short period of time, the entire Earth became a virtual wasteland covered in a thick choking dust cloud! The effect of this was that virtually no sunlight was able to penetrate the cloud, and thus, much of the vegetation that provided food for many of the animals on the Earth soon disappeared! When the vegetation died out, so did the animals that fed on it, which then led to the dying off of the meat-eating carnivores, and on and on and on.

In addition, huge, several *miles* high *worldwide* tsunamis raced away from the impact zone and eventually crashed on the land shores, and then rushed inland inundating large swatches of formerly dry earth, which added greatly to the death and destruction of all of the animals and plants and trees and grasses already taking place throughout the world. One such tsunami raced inland from the Pacific Ocean onto what is now the west coast of the United States, and that tsunami quickly combined with a tsunami that had already raced inland from the impact in the Gulf of Mexico into the southern part of what is also now the United States, where they joined together and flooded the entire western part of the United States all the way inland past the Grand Canyon!

Proof of this exists in what is now the bowl-like valley of Clark County where the city of Las Vegas now exists. Huge quantities of ancient shark and whale bones have been found buried in the soil. And to this day, much of the soil still has so much salinity in it that it is hard to even get grass to grow on it.

In addition to these catastrophes taking place, mega-earthquakes also began erupting all over the world which were caused because the mantle of the Earth was rising and falling like a huge roller coaster ride. This movement caused new mountain ranges to rise up on the surface of the earth, along with other previously existing mountain ranges to collapse.

Finally, since most of the sunlight had been

blocked from reaching the Earth over a long extended period of time *(yom)* because of the huge dust clouds, along with numerous volcanic eruptions exploding all over the world that were flooding the atmosphere with even more choking thick ash clouds, it made the surface of the Earth to drastically cool and the Earth entered into a long extended ice age. Much of North America, including the Great Lakes, was covered by a deep layer of ice.

All of these things combined to cause a massive, worldwide die-off and extinction of most of the deadly creations Satan had prostituted from the *good* Earth that God had originally intended for it to be. Thus, untold millions of these dead creatures eventually got mixed together with untold millions of tons of the dead vegetation in the world where it all eventually got buried deep underground.

Eventually, over many, many millenniums, much of this organic matter got pressed together under great pressure and thus, it coagulated into vast quantities of underground oil deposits. So, thus again, where Satan had planned on great harm by death and destruction by his enormous life forms of deadly beasts, God turned them into good by providing the Earth with the ability for mankind to flourish upon it by the eventual discovery and use of these oil deposits.

With a stunned, speechless, utter horror, followed shortly thereafter by unimaginable anger and murderous hatred that began raging and blasting throughout every fiber and cell of his being, Satan saw what God had done! And with a literal blast of fire bursting from his dragon nostrils, he screamed; *"NO! NO! NO!" This can't be happening! The great I AM has totally destroyed all of my plans again! All of my plans I had for* **"stealing, killing, and destruction"** *(John 10:10) have just gone straight down the crapper!*

*"Why now?" He screamed "Why now, after 180-million years, did the Most High God finally decide to act without any warning? I woke up today just like every other day, then suddenly; all of my plans for evil and destruction were totally wiped out! Just like that! No more plans!*

*"I mean, just think about it! I throw an asteroid against the Earth and the great I AM forms a moon out of it! But He throws an asteroid against the Earth and it kills all of my man-killing dinosaurs, and all of my other precious pets!*

*"It's just not fair! Just not fair. . . Well, oh Great One,*

*you just watch! This war is not over yet! Far from it! I swear to you that it will never end!"*

Several million years later, Satan was still fuming about his defeat with his plan of the huge reptiles to kill and eat God's Man! For all of this time, he had been thinking about what else he could do now to foil God's plan for His Man to rule the Earth.

Then, one day, it came to him. *"Why don't I just go ahead and create a man in my own image and likeness and let them populate the Earth before God creates His Man. Therefore, my Man will become much greater in numbers and thereby more powerful, and then they will be able to overwhelm and kill off any Man God eventually places here on the Earth!"* He thought.

*"I mean, there is nothing stopping me from doing it. Everything I need to create my own Man is already in the Earth because of the Laws of Nature that the great I AM foolishly put into it when He blasted Creation out of His mouth. So, here goes; a Man in my own image and likeness."*

Yes, it was, without any doubt, possible for Satan to bring forth his attempt at a Man upon the Earth. Satan had already proven that he had the ability to cause 'life' to arise upon the Earth by his creation of the dinosaurs and other deadly beasts. But the only thing lacking in Satan's ability to bring forth a Man upon the Earth that would be able to compete with and possibly eventually subdue God's Man, was Satan's lack of intelligence and perfection.

Yes, Satan is powerful. But in comparison to God, well, there is just no comparison, so I'll just let the fossil record speak for itself.

Without any possible rebuttal, the fossil record proves that *'all' of* the attempts by Satan to create a Man anywhere near resembling what he knew God's Man was going to be like, resulted in absolute complete failures! I mean, just take a look at the ridiculous results undeniably proven by the fossil record! And the ultimate proof is the undeniable fact that not a single one of these attempts of Satan to create at least some kind of an intelligent replica of a 'Man' has survived.

Why? I must ask! And don't give me that garbage of 'survival of the fittest' crap, either!

But yet, one item that Satan has been very successful with is the fact that he has gotten almost all of the so-called, elitist intelligentsia, or those that think they are, anyway, to fully believe that *God's true replica of Himself* actually *descended* from these almost comical, complete failures of Satan to populate the world in his own dominant 'man'! And, also to Satan's credit, he has gotten the so-called elitist intelligentsia, to force most all educational teachers and exponents to teach this belief of evolutionary man to the vulnerable, and easily influenced, youth.

So, in this respect, Darwin's theory of 'evolution' was completely wrong! He was mostly correct in his theory of most life 'naturally evolving' upon the Earth. But again, as far as God's replica of Himself descending from one of Satan's imposters, he was incredibly and indelibly wrong!!! Darwin was no more than a duped ignorant stooge of Satan!

But before I go any further, please allow me to take just a quick repose here and speak directly to the Christian Bible-believing Believers. I know that many of you are just about ready to throw this so-called blasphemous book directly into the fire and burn it! I hear you screaming, *"How can you say that evolution is true? How can you even dare to say that Satan is able to create life? Haven't you read* **John, chapter 1, verse 3: "All things were made by him; and without him was not anything** *made that was made."*

Well, I hear you. And, I must admit, you have a very good point.

"Well of course I do! I told you so." You scream!

But wait just a moment, my screaming friend. I said you have a very good point. I didn't say that you have the right, or correct, point, though.

"WHAT?"

That's right. Now, if you will just calm down a little, I'll explain everything to your complete satisfaction, I promise. And if I don't convince you to change your mind, then yes, go ahead and throw this book in the censorious pile of banned blasphemous books! And then I'll willing submit myself to be burned at the fiery stake, too!

Okay, here goes, so open your mind totally, and listen up. First of all, let me emphatically state that John was completely correct in his statement above! *"All things were made by him; and without him was not anything made that was made."*

But what John was speaking of in this scripture is that *"In the beginning when God created CREATION, God had already created everything!* Or have you forgotten that *everything* that is now, or has ever been, was already included inside of that little sliver of Light that God spoke into existence. Nothing else has ever again since that initial little glow of light ever leaked into Creation!

So, John was completely correct if you put his statement in the proper *'context'!* (Remember how we are supposed to study to make ourselves approved by God? So, God did create everything, but He created it all at the very *'instant'* He spoke Creation into existence! Just most of the *'everything'* did not manifest itself as an entity until many *yoms* later!

But I guarantee you, and I really hope you will agree, that God did not create every insect! I mean, all you have to do is just ask yourself, do you really believe that God would have created those tormenting, *disease-carrying* pesky mosquitos?

So now, please allow me to put the final nail in your now wavering belief that Satan could not possibly bring forth life upon the Earth. I'm going to do this by quoting the Bible, if that's okay with you? But, please, please, please, study this to show yourself approved by God, okay?

So here goes.

*Genesis chapter 1, verse 28; "And God blessed them, and God said unto them, Be fruitful, and multiply, and <u>replenish</u> the earth, and <u>subdue</u> it: and have dominion over the fish of the sea, and over the fowl of the air, and over every living thing that moveth upon the earth."*

By putting this scripture in the proper context and studying it so as to make God proud of you, can you think of any other reason why God would tell 'mankind' (His Man) to **replenish the earth**, and then to specifically say to **subdue it,** unless God's Man needed to populate it to such a degree where he was plentiful enough to be able to **'subdue' it?** Listen, why would God tell His Man to subdue the Earth unless there was something there already that needed *subduing?*

Doesn't make sense in any other context, does it?

But God also told Mankind to be fruitful, and multiply, and *replenish* the earth. I've already stated the reason for this that Mankind had to be plentiful enough to overcome the things that needed to be subdued! But to give you scriptural reference for it, read the Book of Genesis in the Bible about where God told Issacs's family to go to Egypt where they would stay there for 400-years. Why? Because, they needed to become **plentiful** enough to be able to possess the land God had promised to Abraham by **subduing** the inhabitants of that land!

The inhabitants of that land were worshippers of false satanically inspired gods, and many of the women in those lands had had intercourse with the so-called 'sons-of-God', who were fallen angels that had followed Satan in his rebellion against God. These offspring produced giants, and they filled the entire area God had given to Abraham and his 'seed'.

So, hopefully, you Christians are still reading this, because if you are, you're about to get some more schooling here real quick. But as a courteous warning to you, let me post this: *Christians beware! Deep water ahead. Science nerds, rejoice. . . . For now.*

There is a very popular show on television. It airs on the History Channel and is called, Ancient Aliens. I often watch this show and am usually amazed at the hosts' insightful knowledge into ancient prehistoric structures that still exist upon the Earth now. It is fascinating for me to hear the hosts of the show explain how the *'only'* possible way any of these structures could have ever been built was by the knowledge and help of an alien race of superior beings that long ago visited the Earth from somewhere inside of the vast reaches of our Universe. In other words, "Ancient Aliens"!

I don't believe that these folks on this show realize just how much actual truth is in their proclamations. In fact, many of their theories and expositions are 'right on'!

But there are a few things I differ with them on. One is their belief that these 'visitors' to our Earth arrived here from some other planet or star system. This is incorrect, although they really did come from somewhere else within the Universe. But these Ancient Aliens originally came from Heaven.

Two is that they believe these 'visitors' were actual physical beings of far superior intellect than ours. And unfortunately, thirdly, they really believe that these 'aliens' were 'good'! In fact, they believe that these 'visitors' were so good that they could actually be called 'gods'!

Well, the easiest one to rebut is the fact that they were, and are still not, good! These so-called ancient alien 'gods' are the demonic fallen angels who followed their master, Satan, in his rebellion against the true 'good' God of the entire Universe. At one time, these demonic fallen angels were good, but they rejected the way they were created and have instead, become powerful, intelligent, demonic entities hell-bent on instilling enough confusion and havoc upon our world in an all-out attempt to eventually bring about our destruction and deaths.

And these so-called 'gods' were not actual physical beings, either, although, just like all supernatural spirit beings, they have retained their supernatural ability given to them by the real Jehovah God at the time of their creation, to be able to assume a prolonged physical appearance of a man. For proof, the Bible has several instances in it of angels appearing in human form. One such example: *Hebrews 13:2* King James Bible; *"Be not forgetful to entertain strangers: for thereby some have entertained angels unawares."*

The Bible proves their ability to change their appearances. This ability did not end when they rebelled! To this day they have retained 'all' of their previous powers and abilities, which makes them very powerful and *deadly* in their continuing war against us!

I wish that the Apostle Paul, when he wrote the above scripture, would have written it more in this manner; *"When you entertain strangers, be very aware that some strangers may be 'angels', of which some may be 'good' angels sent by God to bless and comfort you!*

*But others may be the fallen demonic angels that followed Satan in his rebellion. These kind are out for one thing only, and that is to* **"steal"** *everything precious to you,* **"kill"** *everyone precious to you, including yourself, and finally,* **"destroying"** *everything you have worked your whole life to hold and possess!*

So, as the hosts of Ancient Aliens declare, those so-called *'alien visitors'* really did, and most importantly, still do exist. And as the Bible declares, they were cast down to our Earth. But remember this even if you don't remember anything else here, those *'alien visitors'* are most definitely still here, if you only know how to discern their presence. And their sole purpose and desire is to ensure your ultimate destruction and death!

So, as a word of caution, *DO NOT* try to entertain, or have any contact, with these spirits! They are nothing but EVIL!

So again, being the fact that these fallen angels possess the ability to alter their appearance into the likeness of mortal man, I truly believe that they did exactly that, and therefore, helped Satan's *latest incarnation of 'man' at that time* to build those huge prehistoric monolithic places and monuments! But I challenge you to do some research, and in almost all cases, you will find that the huge monoliths, such as the Pyramids of Egypt, and the Pyramids in Central and South America all were built to either align with stars and constellations, and/or, with the equinoxes of the Sun, and as places of 'worship' to the so-called 'star gods'.

Why? You may ask.

Well, it's really simple. What did Satan once declare to God before he was thrown out of Heaven? "**How you have fallen from heaven, O Lucifer son of the dawn. You have been cast down to the earth, you who once laid low the nations. You said in your heart, "I will ascend to heaven. I will raise my throne above the stars ofGod. I will sit enthroned on the mount of the**

**assembly, on the uttermost heights of the _sacred mountain._ I will ascend above the tops of the mountains. I will make myself like the Most High" _(Isaiah 14:12-14)._**

Satan and his angels drastically want to be 'gods'! They desire to manifest themselves 'above' all that exists and cause all to worship them! So, they helped the deluded people at that time into believing they were sent to the Earth from the 'heavens' to help the people of the Earth to worship the 'gods that sent them down to Earth. They deluded the people into aligning their demonic temples of worship with the supposed places in the heavens of where they supposedly came from. And in many cases, they deluded the deceived people to instill 'human sacrifice' in some form or another into their worship of the 'gods'. Thereby, Satan got another chance to fulfil his main purpose, and that is to kill mankind!

But until he can fulfil his desired purpose, he is adamant about living out his other fantasy; as being god of this world.

So, you may have noticed that I emphasized the words, *"sacred mountain"* in the above scripture. The reason I did that is I believe that 'sacred mountain' is a euphemism for a structure resembling a pyramid.

Now, this is my belief and I have nothing factually to back it up. But I truly believe that in Heaven, God's throne sets on top of a pyramid structure, and this is where God sits and governs the Universe, along with the affairs of 'man'. So, I truly believe that the reason so many of these ancient places Satan and his fallen angels helped the deluded *people* to build was built in the form of pyramids so that Satan can try to imitate the Most High God and His throne room!

Satan is the master imitator. And he is extremely jealous of God. So, since God cast him out of Heaven and down to the Earth, it only makes sense that he would try to imitate the throne of the Most High here on Earth since he has assumed the title of 'god' of this world! So, as 'god', he wants to sit upon his own throne on his own 'sacred mountain' here on Earth.

Remember, Satan was created specifically to be the angel responsible for the throne of God! And he spent his entire existence from the time he was created by God until he was eventually thrown out of Heaven due to his rebellion, right next to God's throne. So, he was intimately familiar with God's throne, its design, and its shape! Thus, my deduction of why there are pyramid structures here on Earth. I believe whole-heartedly that Satan tried his best to imitate God's throne here on Earth!

And there's another thing I want to bring in here. So, let me ask you something; do you remember the story about Noah and the Ark?

Yeah?

Well okay. Do you also remember where the Bible says that the Ark wound up resting at when the water had receded enough for it to?

"Yeah, man. It was upon Mt. Ararat."

That's right. Now, I want you to just think about that for a minute. . . . Okay, now, have you figured it out yet, what I'm getting at?

No? What do you mean, no?

Okay, stop your begging. I'll tell you. You see, I truly believe that God intentionally made the Ark to rest on top of Mt. Ararat with His own Godly man inside of it. It was not an accident that it happened that way. And by doing so, God was getting ready to slap Satan right smack dab in his huge ugly snout!

Again, Satan had declared; *"I will sit enthroned on the mount of the assembly, on the uttermost heights of the sacred mountain. I will ascend above the tops of the mountains."*

My belief is that God made the Ark to rest on top of Mt. Ararat just to prove to Satan that no matter what Satan may ever try or do, that God was still going to retain the 'top' of the 'mountain'! In other words, God was going to retain the 'high places'! By doing so, God proved to Satan that He was the true real God of all Creation, and He would never relinquish His throne on 'top' of the sacred mountain in Heaven, nor on the Earth!!!

So, listen, let me give you one more proof that Satan and his angels have the ability to change their forms! In the Garden of Eden Satan was described as a "Snake" when he deceived Adam and Eve. If you take this verse literally, as so many Bible-thumpers have been indoctrinated to do, it means that Satan had actually turned into a real slimy snake when he approached them.

But here we go again! I know that I'm probably going to make a bunch of closed-minded judgmental Christians furious at me again.

But the truth is the truth!

So, if proven, it must be irrefutably believed! So again, please keep reading until you at least read my explanation.

So, here goes! I don't personally believe that Satan actually had morphed into the form of a snake in the Garden of Eden.

As proof, I offer the actual words straight out of the Bible! In *Ezekiel, chapter 28,* it is written, ***Thus saith the Lord GOD; Thou sealest up the sum, full of wisdom, and perfect in beauty. Thou hast been in Eden the garden of God;*** *every precious stone was thy covering, the sardius, topaz, and the diamond, the beryl, the onyx, and the jasper, the sapphire, the emerald, and the carbuncle, and gold: the workmanship of thy tabrets and of thy pipes was prepared in thee in the day that thou wast created. Thou art the anointed cherub that covereth; and I have set thee so: thou wast upon* ***the holy mountain of God;*** *thou hast walked up and down in the midst of the stones of fire. Thou wast perfect in thy ways from the day that thou wast created, till iniquity was found in thee.*

So, in any court of law, the above scriptures prove without any recourse that Satan did not appear in the form of a snake to Adam and Eve. They say, and I quote, that the **"anointed Cherub"**, (*Satan*) **"was in Eden, the Garden of God."** That scripture places him directly at the scene of the crime!

Now, read again the description given of him, *while he was at the scene of the crime!* He was **"full of wisdom, _perfect in beauty_."** And, **"every precious stone covered his being."** Nowhere does it say that he appeared as a snake. Instead, it is written that he was covered in all of his wealth and beauty!

Listen, in today's parlor, it means that Satan was 'blinged out'!

So, if you believe the Bible to be true, then the above scriptures leave no room for argument! Satan appeared to Eve and Adam just as God had originally created him; as the *"anointed cherub, who was **full of wisdom, perfect in beauty**, with **every precious stone covering his being"**.*

But now, since you have continued to read this, let me take this a step even further. The *'forbidden fruit'* God had warned them not to partake of was not a 'fruit', or anything else at all that could be actually eaten.

But, even so, it could still be *'consumed'*.

Confused?

Let me explain.

If you *'rightly divide'* the scriptures above, they declare that God told them to not *'partake'* of the 'fruit' from the *Tree* of Good and Evil. Nowhere does it say to not 'eat' of it!

"Then what was the fruit?" You ask.

Simple! The 'fruit' was all of the 'bling' covering Satan! Don't you see, the 'bling' contained the **_'root of all evil'_** that lies in the desires of earthly possessions, wealth, and power, more than a person's love for God! (**1st *Timothy, chapter 6, verse 10*: *For the love of money is the* ROOT OF ALL EVIL: *which while some coveted after, they have erred from the faith, and pierced themselves through with many sorrows."*)

I must ask, could there possibly be a more perfect example described in this scripture of what happened to Adam and Eve after they saw the fabulous wealth of Satan? After they *partook* of the *'forbidden fruit'* of the *root of all evil,* they most definitely pierced themselves through with many sorrows for the rest of their natural lives!

Listen, God is definitely not averse to a person acquiring great wealth, or great influence. But, if you put the *desire* for all of that before your love of God, that is where the poisonous lies that came from the mouth of a simulated snake, will eventually bring about your 'spiritual' death!

"So how come Moses described Satan as a 'snake'?" You ask.

Again, it's simple. The *lies* that flowed so smoothly out of Satan's mouth were full of deadly *'poison'!* Just like a snake's mouth contains deadly poison! The poison that spewed from Satan's mouth brought 'death' to Adam and Eve, and thus, to the human race!

So, Moses's description of Satan in the Garden of Eden was just an *allegorical* description of a poisonous snake! But very, very accurate!

Let me explain this a little further. We have to remember where Moses grew up; it was in Egypt. And what was the written language that the Egyptians used at that time?

Yes; hieroglyphics, which was a 'picture' language that described 'words and actions'. So, Moses was very aware of how to use 'pictures' to describe something.

Now, yes, it's true, Moses used Hebrew to write the 'Torah' ~ meaning the first five books of the Bible. But the Hebrew language he wrote in at that time was an ancient archaic type of Hebraic writing that was very similar to a 'picture language'. So, it is very possible that he actually used a 'word' 'picture' of a poisonous snake to describe Satan, even though Ezekiel gave Satan's actually description later on.

Again, Satan's 'lies' were filled with poison! Nothing Satan promised Adam and Eve ever came true other than the fact that they could live out in the world instead of in the Garden of God.

But the riches; the *fruit* that Satan showed them and said they could have? Nada! From that moment on, Adam and Eve lived a life of poverty, earning what food and shelter they had from the sweat on his head and with callouses on his hands!

And Eve? Well, you women don't even want me to go there! Do you?

So, again, if the Bible is to be believed, which it most definitely is, if you **_rightly divide_** it correctly, then Satan didn't really appear to Adam and Eve as a snake; although I do think that he really, really, loves this description of himself.

Want to know why?

Again, just do a little research. If you will, you will find that all down throughout recorded history there have been many factions throughout the entire world who were, and still are, 'serpent worshippers'!

The Egyptian Pharaoh wore a headdress with a serpent 'god' engraved upon it. In the state of Ohio is the world's longest 'serpent mound' carved into the earth! Check it out.

You will also find that on every continent and most countries on those continents, are many other examples of serpent worshippers! And like Ohio, many of them have huge serpent mounds carved into the Earth.

Almost all of the pyramid structures found in the world today have some kind of 'serpent' worship as their main theme. Many Mayan pyramids are even built in such a way that on certain equinoxes of the sun, the shadow of a serpent will rise and descend all the way along the main stairway going up to the top of the pyramid as the sun moves across the sky!

But, if you want to read something I find absolutely amazing and inspiring, read the story in the book of Exodus about the time Moses threw his staff down in front of Pharaoh, and it turned into a 'serpent'.

You may ask, why is that so amazing and inspiring?

It's very simple. What happened right after that?

Yeah, that's right. Pharaoh's own demonically influenced diviners threw their staffs down, too, and they also turned into serpents.

The significance?

Moses's serpent then ate the two diviner's serpents! Swallowed them whole! Then jumped back into Moses's hand and became a staff once again. (This miracle is still seen to this day when you watch a King Snake eat a poisonous Rattlesnake. Just a quick question here. Do you think there is any significance in the fact that the snake that eats the poisonous one is called a '*King Snake*', and the other one is full of poison?) This example is just God showing His Word coming to life everywhere we look if we are really looking for Him!

But again, the significance of all of this?

God was proving to Satan that no matter what kind of snakiest 'poison' he would ever try to defeat God's "Children" with, *well, it just ain't gonna happen,* as we say down here in Arkansas!

And as we also say down here in Arkansas; God will ALWAYS defeat the 'snake'!!!

## CHAPTER EIGHT

## "God Creates Adam and Eve"

For billions of years, God had allowed Satan to basically exercise full control over the Earth and do his own thing. Only at the times that God saw that Satan had stepped over the line in the sand that God had spiritually drawn, would He step in and then turn what Satan had meant for evil back into good, such as turning the dinosaurs into oil. So, for the following several hundreds of thousands of years after God created oil, He had again sat idly by and watched, much of the time with a great deal of humor, at Satan's foolish attempts to create a man that would be at least similar to what he knew God's man was going to be like.

But again, though, as the fossil record testifies, all of his earliest attempts were huge failures, both in mental capacity and physical appearance. Most of them barely had the mental capacity to even gather enough food to barely sustain themselves. And as for physical appearance, well, as you could plainly see, most of them could barely even walk upright on two feet.

And as previously stated, all of these comical failures eventually died out after failing to generate enough progeny to populate a sustaining race, proving once again that survival of the fittest is more than a theory. Yet, evolutionists fervently believe that somehow or other ~ I guess on the same par as the 'cosmic happenstance' just happening theory, *LOL* ~ a new style and semblance of a little better 'model' of a two-footed creature just 'evolved' out of the previous complete failure.

And if you can believe that, I have a huge bridge very similar to the Golden Gate Bridge located in Arizona I'll sell you really cheap, man!

I mean, will someone please, please, please, with jelly and sugar on it, tell me how can a new and better model just descend out of a decrepit piece of pure ugliness? I mean, these really ugly creatures were not made by General Motors, or even Ford, where an ugly Edsel could eventually be turned into a beautiful Mustang!

Man oh Man!

But yet, there really are a tremendous

number of so-called elitist Intelligentsia comrades who will demand that you who dare to question their beliefs be exiled to the garbage pile of human stupidity forever! . . . By the way, I am gladly and proudly sitting just fine and dandy at the very pinnacle of this garbage heap, because I refute their beliefs 100% percent!

Anyway, after hundreds of thousands of more years went creeping by, and after several more supposedly new and improved different weird incarnations of a cross between man and beast came about, and then died out, Satan's imitations of God's coming 'Man' were getting somewhat closer to fruition in mental abilities and appearance. In fact, the latest imitations were actually able to walk mostly upright even though they were still as ugly as sin!

But, of these new and supposedly improved 'models' fresh off the satanic assembly line, some of them were even able to make rudimentary tools out of rocks to help in their foraging.

But then again, so can Crows!

Anyway, two of Satan's latest attempts would much later on be called, Cro-Magnon man, and **Neanderthals.** Modern man has cataloged these two species to fall into the 'homo' biological classification of which they have also classified modern man as such, so they spout the belief that these ugly as sin creatures were our nearest ugly cousins in our dynamic *evolutionary* march upwards to finally reach our 'beautiful' esteemed stature we have arrived at today.

But, unfortunately, both of these species also died out roughly around 40,000 thousand years ago or so, again after not being able to produce enough progeny to establish dominance over the Earth.

But one thing you can grudgingly say about Satan, he has the stamina of a bull, and about the same intelligence as one in a China shop, because he just doesn't quit! So finally, after many more years had gone by, Satan was able to produce a much closer imitation, at least in appearance, of what he knew God's Man was going to be.

In fact, the only thing that would really separate Satan's Man from God's Man would be by what was in their hearts and minds. Satan's Man would have a cold hard heart where he would invariably want to imitate his own evil facilitator by trying to attain as much power over his fellow man as possible by starting wars and conflicts that always resulted in huge instances of bloody, gory deaths! Then, the victors would acquire complete control over any survivors and enslave them. Therefore, his stated desire to *'steal, kill, and destroy'* would be accomplished on a wholesale basis.

As for his mind, Satan assumed complete control over his Man's reprobate mind. Thereby, he was able to turn his Man into a captive slave of his own desires. He was also able to easily manipulate his Man into creating and worshipping multitudes of false 'gods', all of course, designed to worship Satan's own image and likeness in some way or another.

He, also, infused the carnal desire within them to have intercourse with just about anything that was alive, including both the Cro-Magnons and Neanderthals before they died out. Not only does some of the DNA from these demonically created species still remain in large quantities of individuals today, but their unnatural desire to have sex with any and everything also remains in these same demonically controlled individuals today.

Example, just take a casual look into the lives of the Hollywood so-called 'stars', along with the lives of most of the musical so-called 'stars' of today. Rabbits have nothing on these people!

An even more perfect example of this is in the deviant sexual escapades of some former Presidents of the USA; our so-called esteemed leaders. In fact, it's been claimed that one such demonically controlled former president would have screwed a living snake if nothing else was around! And I won't even pay any heed at all to those of you wannabe comedians out there who are loudly proclaiming that he already had because he was married to one.

Now hush!!

But after finally succeeding in his attempts to fairly accurately make an imitation copy of what God's Man was going to look like on the surface, Satan decided to do what he thought would completely solidify his ownership and control of the Earth by a very shrewd method. He told his demonic angels to change their appearances into what his own Man looked like.

Then, he succeeded in having his demonic angels manipulate *true reality* either by an 'alternate reality' with the use of **hallucinogenic** drugs or, by the use of some form of 'virtual reality' or, some other form of deceiving magical tricks that easily deceived the eyes and minds of his Man. This was so he could cause their manipulated eyes and minds to believe that they were seeing his demonic angels descending down from the sky in flaming vessels of fire so they could claim to be "sons of God". And because of this satanic deception, and of the handsome appearances of the so-called sons of God, they easily were able to seduce the female populations of Satan's imitations. (See Genesis 6 for scriptural reference, plus many other historical and oral references of Ancient Astronauts and also of Giants living in the lands.)

Thus, many of the offspring of the illicit unions between the sons of god and the daughters of men were where the 'giants' of old came from. If you were to do an in-depth study of ancient history, you will find that almost all *pre-adamic cultures* have stories passed down from generation to generation about giants having lived in their lands. And, I truly believe that Satan was able to instill in these creatures the great knowledge and superhuman strength needed to build towering prehistoric structures such as the Pyramids and Sphinx of Egypt, along with all of the other huge complex prehistoric structures found around the world that fall into this category.

But, learned scholars have also found a lot of corroborating evidence that prove that many of these structures were places of worship to their heathen false gods that Satan had deceived them into worshipping. Also, in many of these places evidence has been found to show that human sacrifice had also taken place. This should not come as any surprise because to Satan's complete and utter joy, human sacrifice has been practiced all down throughout history.

But to what should be *everyone's* utter and complete shame, the most successful period of time in all of history in the barbaric heathen practice of human sacrifice in the form of infanticide has been during the twentieth and twenty-first centuries! During the early part of the twentieth-century, a demonically influenced and controlled woman by the name of Margaret Sanger founded several organizations that preached birth control for women.

Absolutely nothing wrong with that. But one of the organizations she founded eventually became Planned Parenthood, which undoubtable is one of the largest perpetrators of infanticide in the history of the world! Just since abortion was legalized in the United States in 1973, Planned Parenthood has been totally responsible for the *murder* of close to **8–MILLION BABIES!!!** Of which the vast majority of these were of black, or African American babies, since Margaret Sanger was an evil, evil racist who believed absolutely in the practice of 'eugenics', and also of population control whereby survival of the fittest is thrown out the door and replaced by a 'planned government population control' method!

Eugenics is the satanic belief and practice that a population can be drastically improved by "selective breeding"! Thus, the hugely 'racist' Sanger believed that by eliminating the **black population** through 'birth control' *by any means possible,* the white population would drastically improve its stature. It is, therefore, very easy to place Sanger, and thus, Planned Parenthood, at the very top of the vile evilness of the white supremacy establishment!

Listen up, another name for eugenics is *'ethnic cleansing',* which is the evil practice of killing every person that is of a particular race by a supposedly superior race, such as the Nazi's evil attempt to kill all of the Jews in the world during World War II, and the Armenian Holocaust that took place during the early part of the twentieth century that was perpetrated by the Ottoman Empire, that later on descended down to become the country of Turkey today!

But, by Margaret Sanger's frantically dedicated and evil use of *ethnic cleansing* practiced through her evil, satanically inspired progeny of Planned Parenthood, along with many other baby-killing companies that followed directly in the same footsteps of Planned Parenthood that were created specifically for the purpose of splattering the torn-apart remains of human babies all over their blood-drenched satanic altars, all of these combined have killed almost ***18-MILLION BLACK BABIES*** just since 1973 in the United States alone!!! These 18-million ***stolen, killed, and destroyed*** dead black babies would equal to 44% percent of today's total 'black' population!

Let me put this another way. *Almost one-half* of the total black African-American population of people living at this time has been brutally ***'murdered'*** here in the United States just since 1973! This is NOT police brutality, people! This is Planned Parenthood and all other baby-killing companies brutality in its basest form of population control by the use of eugenics!

But the United States is not the only

unrepentant killer of human babies in the world. Just since 1980, just **40-years,** almost ***ONE AND A HALF BILLION BABIES of all colors and races have been slaughtered like little more than cattle worldwide!!!***

God, please forgive us!!!!

Okay, I'll get off my soapbox now and move on. I can hear you screaming about another item I mentioned above.

Now, I know that most, if not all of you religiously righteous, Bible-thumping believing Believers are just about to have a coronary attack because of me using the word, *"pre-adamic"* above. But before you actually burn good ole blasphemous me at the stake, let me remind you again of ***2nd Timothy 2:15; "Study to show yourself approved by God, a workman that needed not be ashamed, 'rightly dividing' the word of truth"***!

Oh my! Did I just hear a bunch of you screaming, ***"OH NO! HE'S NOT GOING TO DO IT TO US AGAIN, IS HE?"***

Yep, you bet I am, my friend.

Okay, the first scripture I am going to rightly divide, and prove as absolutely 100% correct, is concerning about there being giants in the Earth. The first scripture is **Genesis 6:4; _There were giants in the earth in those days;_ _and also after that, when the sons of God came in unto the daughters of men, and they bare children to them, the same became mighty men which were of old, men of renown._**

**_Numbers 13:33 - And there we saw the giants, the sons of Anak, which come of the giants: and we were in our own sight as grasshoppers, and so we were in their sight._**

**_Deuteronomy 1:28 - Whither shall we go up? Our brethren have discouraged our heart, saying, The people is greater and taller than we; the cities are great and walled up to heaven; and moreover we have seen the sons of the Anakims there._**

**_1st Samuel 17:4-7 And there went out a champion out of the camp of the Philistines, named Goliath, of Gath, whose height was six cubits and a span. ⁵ And he had an helmet of brass upon his head, and he was armed with a coat of mail; and the weight of the coat was five thousand shekels of brass. ⁶ And he had greaves_**

*of brass upon his legs, and a target of brass between his shoulders. [7] And the staff of his spear was like a weaver's beam; and his spear's head weighed six hundred shekels of iron: and one bearing a shield went before him.*

*Deuteronomy 3:11 - For only Og king of Bashan remained of the remnant of giants; behold, his bedstead was a bedstead of iron; is it not in Rabbath of the children of Ammon? nine cubits was the length thereof, and four cubits the breadth of it, after the cubit of a man.*

---

And there are several other scriptures I did not take the *"yom"* to list here but they're in the Bible if you will just take the *'yom'* to look for. So, if you truly believe that the Bible is correct, then you must believe that giants once existed upon the Earth; many, many giants! Many different races of giants. All over the earth!

Perhaps the most famous example of all though, was the clash between David and the giant Goliath. But did you know that Goliath had four brothers who were also giants? To find out, rightly divide the word of truth in 2nd Samuel 21:18-22

But I hear you still screaming; "Yeah, man! We already knew that there were giants in the world. You ain't telling us nothing we didn't already know."

Okay, I knew that you already knew about giants.

But most of all, my friend, I just wanted to get your attention finely attuned and to allow you time to put on your life-preservers, again. You see, we're about to jump into some really deep water right now in our glorious trip through the Bible in our attempt in rightly dividing biblical scripture into the *word of truth!*

So, let's begin right at the beginning of the Bible, okay?

Throughout Genesis, Chapter 1, the Bible details all the things and creatures that God made. *And in every instance,* afterwards, God said that it **"was good".**

Now, unless you want to flat out call God a liar, then there is no way that He could have created poisonous snakes and spiders and scorpions and all of the other nasty, *man-killing* things that exist in the world. He, especially, could not have been the creator of the blood-thirsty dinosaurs, and the other blood-drinking creatures that occupied that *"yom"* in the Earth's history.

But yet, there is absolutely no doubt that those things existed. And, especially, there is no doubt that nasty, man-killing snakes and spiders, and scorpions, and such still exist to this day. As such, there should be no doubts in anyone's mind that these things cannot possibly be called *"good"*.

So, that leaves just one determination. All of these nasty creatures, plus all of the other scary and dangerous creatures I haven't mentioned, could not possibly have been made by God!

Even so, I know that some of you are still going to say that all of those things all play a huge part in controlling 'nature'. You're saying that without them, nature would explode out of control!

And of course, you are absolutely correct. But get that huge grin off your face!

Just in case you may have forgotten, let me remind you that the Garden of Eden was a place of "life" where everything in it was *'good'!* Inside of the Garden was the *'Tree of Life'!* Therefore, *death* did not exist inside of the Garden!

So, by default, there could not be any poisonous creatures of any kind. Nor could there be even a nasty biting mosquito that could have existed in the Garden of Eden. There could not be any *evil* at all of any kind *ever,* until one day, a poisonous bling-covered 'snake' snuck into the Garden and planted himself smack dab in front of the Tree of Life, thereby in effect, blinding Adam and Eve's eyes from the one item that provided Life to them!

But, the Garden of Eden was the one, *and only,* place on Earth where God made everything "good". So, to quell the argument you're about to present, let me just say lest you forgot, that Satan had been *"cast down to the Earth"* as the *Word of Truth* declares. So, he had full control over it *outside of the Garden of Eden.*

That, my friend, was the entire reason in the first place that God made a special place of *sanctuary* upon the Earth called the Garden of Eden!!! It was so He could *'meet all of the needs' of His Man* without the death and destruction going on all around in the rest of the world!

Listen, it is just plain old common sense that if the rest of the Earth had not been under Satan's control, then God would have just made the whole entire world into a Garden of Eden, and then, everything that *had ever existed* upon it would have been 'good'!

But it wasn't!! We 'know' that!

So, I proclaim at the top of my voice the undeniable *fact* that any and all creatures of any kind that had any kind of *evil* intentions were not "good" when they were created! So, unless you're just a die-hard blind idiot that can't see his own face in a lighted mirror, you must admit that some other *force of darkness* had manipulated the "Good Earth" that God had intended the whole Earth to be.

And if you don't believe that, just pull the straps on your life-preserver a little tighter. Because, I'm going to prove it to you by *rightly dividing the Word of Truth* for you, *again!*

Now, let's go to **Genesis chapter 2:7-9 And the LORD God formed man _of the dust of the_** *ground, and breathed into his nostrils the breath of life; and man became a living soul. [8] And the L<small>ORD</small> God planted a garden eastward in Eden; and there he put the man whom he had formed. [9] And <u>out of the ground</u> made the L<small>ORD</small> God to grow every tree that is pleasant to the sight, and good for food; the tree of life also in the midst of the garden, and the tree of knowledge of good and evil.*

*Genesis chapter 2: 19 And <u>out of the ground</u> the L<small>ORD</small> God formed every beast of the field, and every fowl of the air; and brought them unto Adam to see what he would call them: and whatsoever Adam called every living creature, that was the name thereof.*

**N**ow, I want to stop right here because I'm going to prove, once and for all, that in the original Light that exploded out in the Big Bang, that God had already included *'everything'* that would ever exist in the Universe, and especially, everything that would ever be included in 'nature' on the Earth for 'Life' to 'naturally' arise upon it. Look at the scripture above, it says that ***God formed man <u>out of the dust of the ground!</u>***

Yeah, man! God didn't 'speak' Man into

existence like He had done with the Big Bang. NO! He spoke Creation into existence only 'once'! Therefore, God used the material **He had already created** in the Big Bang and then placed in nature in the Earth to form Man out of!

Verse 19 declares that **"out of the ground"** God formed every beast and fowl! God also formed all of them out of what He had ALREADY made as an original part of the Earth! In 'fact' 'all' life that has ever arisen upon the Earth, came about because of the elements that God 'had already' placed in the Earth at the *"yom"* of the Big Bang!

## Case closed! *SLAMMED SHUT!!!*

Now, I can hear you screaming that the scripture above says God created *"every"* beast and fowl! And, of course, you're correct again.

But I again declare the undeniable argument that God *only* made the things that were "good", because He, himself, declared it so! Nowhere does it say that God also made things that were evil! But yet, evil things undeniably surely existed and still do exist!

You see, if you *rightly divide* the scriptures above, it plainly says that God formed all of the beasts and fowls *'while He was in the Garden of Eden'* because, that is the place where God brought them before Adam to name them! Use plain old commonsense and it will tell you that if God had made *all* creatures, including the mean *evil* ones, then He would have had to bring 'evil' into the Garden of Eden for Adam to name where *everything* was only GOOD! Commonsense should tell you that God would never have allowed anything other than GOOD to ever step foot inside of His *Paradise* on Earth!

Perfect example. There was only one time that Evil ever entered into the Garden of Eden. That was when Satan entered into it and he used his time there to seduce Adam and Eve.

So, let's see what happened right after that.

Yeah, that's right. God kicked the Evil in the form of Satan out of His Garden, along with the evil that now had taken up permanent residence inside of the Man that God had made.

Don't you understand, the Garden of Eden was a perfect metaphor of Heaven! God created it to be a little slice of *"Heaven"* on Earth! Since no evil can ever be found in Heaven, likewise, God was not going to allow evil to exist in His own earthly replica of His own home.

So, again, God did not create any creature of any kind that had evil intentions of any sort in mind! But Satan sure did!

But wait, there's more. The Bible says that Adam lived for 930 years. *But is that correct?* Or do we need to *rightly divide* some more *words of truth?*

# CHAPTER NINE

## "How Long Did Adam *'REALLY'* Live"

T he Bible specifically states that Adam lived for 930 years. But is this really true?

Well, yes, if you have believed every preacher and Bible teacher you've ever listened to. But wait, because I'm about to do it to you again! I'm going to blow their beliefs and statements right out of the water!

*"But WAIT!"* You scream! "The *BIBLE* specifically says that Adam lived for 930 years. So, are you going to call the Bible a liar?"

No, my friend, I'm not that stupid! But I would have thought by now that you would have caught on to the fact that many times you have to *"rightly divide the word of truth"*. Many times what your eyes see and what your brain assimilates is not the whole story. So, at those times, you have to *'study to show yourself approved by God, a workman that needs not be ashamed'* to see and understand the whole meaning.

So, let's read some *words of truth,* shall we?

Let's actually start in the Bible at the point in time when God created Adam, and then go forward from there. But I must warn you, keep that life preserver pulled tight, okay? Along our journey, the water is going to get deep at times.

***Genesis 1:26-28 And God said, Let us make man in our image, after our likeness: and let them have dominion over the fish of the sea, and over the fowl of the air, and over the cattle, and <u>over all the earth,</u> and over every creeping thing that creepeth upon the earth.***

(Side Note: By God giving 'Mankind' authority ***"over every creeping thing that creepeth upon the earth",*** God was unequivocally giving Mankind full authority over the creeping 'snake' that had deceived Adam and Eve in the Garden.)

***So God created man in his own image, in the image of God created he him; male and female created he them. <u>And God blessed them, and God said unto them, Be fruitful, and multiply, and replenish the earth, and subdue it:</u> and have dominion over the fish of the sea, and over the***

*fowl of the air, and over every living thing that moveth upon the earth.*

*Genesis 5:1-5 This is the book of the generations of Adam. In the 'day' (yom) that God created man, in the likeness of God made he him; 2 Male and female created he them; and blessed them, and called their name Adam, in the day when they were created. 3 And Adam lived an hundred and thirty years, and begat a son in his own likeness, after his image; and called his name Seth: 4 And the days of Adam after he had begotten Seth were eight hundred years: and he begat sons and daughters: 5 And all the days that Adam lived were nine hundred and thirty years: and he died.*

Okay, so again, I ask, was 930 years Adam's real age?

Well, yes!

And *NO!*

"Say what?" You scream!

Well, yes, 930 years is the time Adam lived according to the Bible, and probably every preacher you have ever heard. And, of course, they and the Bible are correct, *as far as it goes.*

But, that's not Adam's 'true *real*' age. So, get those life preserver straps pulled tight, again, okay? Now, let's go *"rightly divide"* some more *"words of truth"*, shall we?

Now, the first verse above says it's going to give the details about the generations of Adam. But, unfortunately, it again contains the word translated, "day", in the King James Version. And again, it should not have been! The word *"yom"* should have been translated more loosely as, *"In the **time period** that God created man";* and so on.

Anyway, let's skip on down to verse 3. It says that Adam lived a hundred and thirty years and had a son he named Seth. Now, I'm not going to be real nit-picky right here, because this is technically true, but then again, it's not *completely* true, either.

You see, this verse completely ignores both of Adam's previous two sons that were previously mentioned in the Bible, Cain and Abel. I'm sure the reason for that is because the "good" son, Abel, was murdered by the 'bad' son, Cain. So, I'm not going to get off track and follow this very important track right here just yet. I will a little later on, though.

So, let's get back to our original subject. Again, the above chapter is the 5th chapter in the book of Genesis. So, the *time period* (yom) in between Chapter 2, verse 7, when God created Adam, and Chapter 5, when it begins giving Adam's age, is a lot of fabulous knowledge located just under the surface. *Or, out in the deep water!*

Let's review it a little, shall we?

Chapter 2:7, God creates Adam and Eve. And what does He tell them to do?

Yeah, that's right; to put it bluntly; God told them to have a lot of sex so they can multiply themselves many times over. This is why I highlighted the word **'their'** in verse 2 in the above scripture where it says that God called *'their'* name, Adam. So what God was actually telling the very *'first'* 'Adam' and 'Eve' to do was to have intercourse and produce many more 'Adams' after their own kind.

In other words, God personally created only one Adam and one Eve, and that *is all* He put His hands to do. Then, the rest was up to Adam and Eve to produce more Adams' ~ or more specifically, more of *'mankind'* whom God had named as Adam. Remember the *"laws of nature"* God had placed into His creation at the very beginning of Creation?

But why did God tell them to produce many more Adams'?

Yep, you're right again. It was so they could eventually produce enough Adams' to be able to *subdue* all of Satan's evil replicas of himself already existing on the Earth, and then have enough Adams' to *replenish* it with God's people.

So, let me ask you something; how many additional Adams' do you think it would take for them to be able to successfully subdue the entire Earth and then replenish it? Because, if you only include Cain, Abel, Seth, and the other sons and daughters Adam and Eve produced after they were kicked out of the Garden of Eden, they would have actually failed miserably in doing what God had commanded them to do.

So, let me ask you again, how many "Adams'
would it have taken for them to successfully
subdue the entire Earth and then replenish it?

What do you mean you don't know?

Okay, I'll cut you some slack. But, let me ask,
do you think 10 replicas could have subdued the
Earth?

"No way?" You say.

Then what about a hundred? . . . . . . A
thousand? . . . . . A hundred thousand?

Well, let me check your biblical knowledge a
little then with a perfect example of how many
'Adams' were needed to subdue and replenish a
certain land later on in history. Do you remember
the history about how God called Abraham out of
his home town of Haran in southern Iraq and then
brought him to the land of Canaan? And about how
God promised to give all of that land to Abraham
and his seed? *Genesis 13*

Good. Glad you remember all of that. But, do
you know why Abraham never personally lived to
see that promise come true during his lifetime?

No?

Well, I'll tell you. It was because during the *"yom"* of Abraham's life he never produced enough *'Adams'* to be able to successfully *'subdue'* all of Satan's replicas of himself already living in that land. And even if he had been able to *subdue* it, he still would not have had the time to reproduce enough *'Adams'* to be able to successfully *replenish* the land with his own seed.

Do you remember the story about the giants?

Well, besides that land having thousands of Satan's imitations of God's Man already living in the land, numerous races of giants also lived in that land, too! Later on, Abraham's descendants would come face to face with Goliath and his brothers in that land, plus many, many more giants not specifically mentioned by name!

So, even by the *"yom"* of Abraham's grandson, Jacob ~ renamed Israel ~ and Israel's twelve sons, and their progeny, they still had not produced enough individuals to be able to subdue the land and also to replenish it. Altogether, they counted a total of only around seventy individuals. *(Remember this: Three generations of God's righteous replicas of Himself descended from Abraham were then living, and yet, they had only produced around seventy members of their family.)*

So, what did God do to help Abraham's progeny to be able to eventually subdue the land so they could replenish it with their own family members?

Yep, you're right. As He often has done all throughout history, God manipulated nature to accomplish His desires and plans. He allowed a famine to come upon the land. In effect; because of this famine He caused the seventy members of Abraham's progeny to go down to Egypt where they would stay for the next 400 years.

Look closely here, what appeared in the natural to these seventy individuals to be an awful, life robbing disaster of eminent death by thirst and starvation to these people, was actually God manipulating their reality to get the people to do His will, which was to move them out to a much better place at that time where there was plenty of food and land for them.

I have to take just a minute here and ask, have you ever experienced what appeared in the natural to be an eminent disaster coming upon you? Maybe a job loss, or an unseen divorce, or anything else that keeps you up at night? Well, let me assure you that whatever it is, God is just waiting for you to be fully willing to submit your will completely to Him no matter what it looks like to you in the natural, and then He will, without a doubt, move you up and into something so much better than you ever expected or even dared to dream about!

But, you may ask why God made them stay in Egypt for 400 years, and eventually, be cruelly enslaved instead of just making them stay there just until the famine was over?

It's very simple.

During that *"yom"* of four-hundred years God's chosen people had a lot of sex which eventually resulted in them multiplying themselves from around an initial 70 people to somewhere between 2-million people upwards to 3 ½-million people depending on who you believe. But, either way, by the end of those four-hundred years of *yom*, there certainly were enough descendants of Abraham produced to be able to *subdue* the entire land God had promised to give to Abraham, and then to *replenish* it.

You see, here's God's promise He gave to Abraham. *Genesis 13:14 And the LORD said unto Abram, after that Lot was separated from him, Lift up now thine eyes, and look from the place where thou art northward, and southward, and eastward, and westward: For all the land which thou seest, to thee will I give it, and to thy seed forever. <u>And I will make thy seed as the dust of the earth: so that if a man can number the dust of the earth, then shall thy seed also be numbered.</u>*

*Also Genesis 15:18 On that day the Lord made a covenant with Abram and said, "To your descendants I give this land, from the Wadi[e] of Egypt to the great river, the Euphrates— 19 the land of the Kenites, Kenizzites, Kadmonites, 20 Hittites, Perizzites, Rephaites, 21 Amorites, Canaanites, Girgashites and Jebusites."*

So, during those 400 years in Egypt, God's promise to Abraham was greatly fulfilled because either 2-million people, or 3 ½-million people, would certainly have *stirred up enough dust to cover the land* as they traversed across the Arabian desert. With a dust cloud so huge that it appeared that the entire Arabian Desert was covered by an invading horde to the current people living in Abraham's land, is it any wonder why Rahab was so eager to help the spies? *(Joshua 2:1-7)*

She definitely knew without a doubt which side her bread was buttered on!

Anyway, after 400-years had gone by, there certainly were enough of God's holy and righteous replicas of Himself to move back into the land God had given to Abraham, and to be able to fully *'subdue' it!* And they certainly would have been able to fully subdue it *if only they would have believed God and followed His instructions to them to the full-letter* instead of getting weary of the fight and quitting before the war was fully won!

In other words, if only they would have kept the 'good' in themselves that God had imbued in them and fully followed His instructions to them, they could have eventually *subdued* all of the evil satanic ones living in that land. And if they had followed God's instructions to the letter as He had told them to do, I venture to say that most of the problems that exist now in our "yom" of time in that area of the Earth, would not exist now!

Anyway, getting back to Adam and Eve. So over a *"yom"* of 400 years, God's people went from 70 people to at least 2-million people. Using this as an example let me ask, how long do you think it would have taken Adam and Eve, and their progeny, to produce enough replicas of themselves to be able to subdue the *entire* Earth?

Yeah. Probably a long, long *"yom"*, huh? If it took 400-years for Abraham to have enough descendants to subdue and replenish just a relatively small area of the Earth, then undoubtedly, it probably would have taken Adam and Eve, and their progeny, many thousands of years before God would have said there were enough of them to go forth and subdue the entire Earth and then replenish it!

Don't you think?

And remember this, too. During the time *(yom)* that all of these 'Adams' were living inside of the Garden of Eden, neither the original Eve, nor any other of the female *'Adams'* suffered any pain during childbirth.

Yeah! That's right. None of the females living in the Garden of Eden suffered any pain at all during childbirth. It was not until after they sinned that pain during childbirth happened.

*Read it. It's in the Bible.*

So, folks, that fact alone makes me believe that Adam and Eve, and eventually, their progeny, did exactly what God had commanded them to do; and that was to have lots and lots and lots of sex, and thereby, multiplying themselves many, many times over, *while still living in the Garden of Eden.*

Yes, I know that the Bible only mentions three children of Adam and Eve by name, and then briefly mentions that they had other sons and daughters following Seth. But in actuality, this leaves a vast conundrum. Remember, God had specifically ordered them to *"be fruitful, fill the earth, and subdue it."*

So, Bubba and Mrs. Bubba, here is where the water gets a little deep. If Cain, Abel, Seth, and these other children were the only children Adam and Eve had, then Adam and Eve had failed miserably to uphold the commands of God to *be fruitful, fill the earth, and subdue it.* They might have been somewhat fruitful, but they were a long,

long way from being able to fill the earth with enough progeny to subdue and replenish it.

So, what possible solution is there for this conundrum?

Well listen; there is only one possible answer here. It is that they did do exactly as God had commanded them to do, and they and their offspring produced thousands and thousands of more offspring *while they were still living in the Garden of Eden!*

Again, that is the *ONLY* possibility!!!

So, listen closely here and I'll explain how you can see this to be true. The Bible clearly declares that God came down and took long walks with Adam on a daily basis, so that leaves no choice but to believe that Adam and Eve lived in the Garden of Eden for a very long time. I mean, how long do you think it would possibly have taken Adam to just name all of the 'good' living things on the Earth?

Well, an answer to that is that mankind has still not named every living creature to this day. So, to me, it is very plain to see that when Adam wasn't naming living creatures, he was probably very happy to be fulfilling God's commandment to him by being fruitful with Eve!

Listen, big Bubba, here's another glorious fact that you may have overlooked; Eve was *'naked'*, man! And she was, and still is, the most beautiful woman ever created because she, too, was made in the likeness and image of God. So, are you actually going to try to tell me that Adam wasn't jumping her bones every chance he got?

Come on, man! Who are you if you don't believe that?!!!

So anyway, I'm going to wrap this up right now with no further argument, so keep those straps pulled really tight.

I believe that Adam and Eve produced lots and lots of children while still living in the Garden of Eden before they eventually sinned and were kicked out of the Garden. Now, each child they produced took nine months before it popped out. So, my friend, for them to produce enough progeny to be able to fill the entire earth and to subdue it, they would have had to live in the Garden of Eden for a very, very long *'yom'*. Therefore, the original

Adam *'actually'* lived for a whole lot longer than 930 years! You see, 930 years was *only* the amount of *'yom'* <u>after he got kicked out of the Garden of Eden.</u>

You see, up until the *'yom'* that Adam and Eve sinned, they were actually immortal beings! Listen, they were made 'good', so there was nothing 'bad' that could harm them in any way whatsoever! They also lived in a place that was only 'good', which was an earthly replica of Heaven ~ a perfect metaphor as such! So, there was nothing 'bad' that could be in their little slice of Heaven on Earth.

Also, this perfect metaphor of Heaven contained the Tree of Life. That meant that every time they partook of it, it instilled 'life' into them!

So, here it is again, 'death' only entered into them once they had partaken of the 'forbidden fruit'; which was their desire for the so-called 'bling' of the world Satan had shown to them. As such, Adam's 930 years were only measured from the time that death entered into him at the time of his sin! So, Adam's *true* age is unknown, but it was sure a whole lot longer than 930 years.

In fact, science tells us that 'modern' humans arose upon the Earth approximately 40,000 years ago. I agree with this, but I must modify the science statement to say that approximately 40,000 years ago (or a long yom ago) God created the original Adam and Eve, and they produced many, many offspring while living in the Garden of Eden, and their offspring produced many, many offspring, and so on and so on before they sinned and got kicked out of it.

So, the next question I assume you're going to ask me is; where exactly was the Garden of Eden located? And just how large could it possibly have been to be able to feed and support literally thousands and thousands of 'Adams'?

And what about Cain and Abel?

# CHAPTER TEN

## "Where Was the Garden of Eden Located, and What About Cain and Abel"

The Bible gives us some very good clues about where the location of the Garden of Eden was once located. It says that the Garden was watered by four rivers, one of which is still using the same name as when Moses wrote the story. That is the Euphrates River. But the other three are kind of mysteries. That is if you don't do your job of *studying to show yourself approved by God.*

Many historians, archeologists, and other supposedly learned Biblical scholars have speculated that the Tigris River was another one of the four rivers. And very probably, in this case, they are correct. Unfortunately, though, some of these same scholars have also postulated that since both of these rivers empty into the Persian Gulf, and since the water level in that gulf was much lower a long yom ago than it is now, then the other two rivers could possibly be located under the Gulf waters now. Satellite photos have also given some credence to this theory.

But, let's look at the scriptures again, and then do a little rightly dividing, okay?

**Genesis 2:8-14** *⁸ And the L*ORD *God planted a garden eastward in Eden; and there he put the man whom he had formed. ⁹ And out of the ground made the L*ORD *God to grow every tree that is pleasant to the sight, and good for food; the tree of life also in the midst of the garden, and the tree of knowledge of good and evil.*

*¹⁰ And a river went out of Eden to water the garden; and from thence it was parted, and became into four heads. ¹¹ The name of the first is Pison: that is it which compasseth the whole land of Havilah, where there is gold; ¹² And the gold of that land is good: there is bdellium and the onyx stone.*

*¹³ And the name of the second river is Gihon: the same is it that compasseth the whole land of Ethiopia. ¹⁴ And the name of the third river is Hiddekel: that is it which goeth toward the east of Assyria. And the fourth river is Euphrates.*

If you really study these scriptures, they actually give the exact location of the Garden of Eden. But, as for me, all I have ever heard from anyone, including scholars and backyard amateur historians, is them basing their location for the Garden of Eden only on the *rivers* mentioned above. And since the Euphrates is the only one anyone *thinks* they know exactly where it is, most, if not all, of them say the Garden of Eden was in what is now southern Iraq.

But if the so-called experts are wrong, just where might the Garden of Eden have been really located?

Well, there are a lot more clues in the above scriptures if you just look close enough. Besides the mention of rivers, they also give us land, gold, other minerals, a place called Havilah, and even the name of a present-day country, Ethiopia.

"Ethiopia! Wow! That's a really long way from southern Iraq!" You scream!

Anyway, here's my hypothesis, and take it as only that. I am not saying this was the actual location because no one knows for sure. But I really do believe that the scriptures ~ and historical records, including historic climatic records, and especially, present-day geologic formations ~ give us a great hint at where it was, and that my postulation is the correct one.

With that being said, let's begin.

We know the Euphrates River was one of the rivers that watered the Garden. From there, based on this scripture; ***And the name of the third river is Hiddekel: that is it which goeth toward the <u>east of Assyria</u>,*** I believe whole-heartedly that this scripture is describing the Tigris River. We know where ancient Assyria was located. It occupied part of present-day Syria, and then goes eastward into Iraq and even Iran, all of which are ***"east of Assyria".***

So, based on this information, I believe that the *northern boundary* of the Garden actually began in southern Turkey just south of the ruins of an ancient pre-historical place now called, Gobekli Tepe. (I will tell you more about Gobekli Tepe a little later on.) But this is the place where the headwaters of the Euphrates River begin as it flows southeast towards the Persian Gulf.

So, we now know the location of two of the

rivers that watered the Garden, and based on them, we also now have the northern border of it.

Now, let's take the country of Havilah, and the gold and the other minerals as another clue. Let's start with the gold. In the Bible in *1 Kings 9:28* **it says, *And they came to Ophir, and fetched from thence gold, four hundred and twenty talents, and brought it to king Solomon.***

Ophir is generally accepted as the place where Solomon's gold mines were located. It was bordered on the northeast side of the Red Sea. On present day maps, it would now be a part of southeastern Saudi Arabia and the northwest side of southern Yemen.

It is also generally accepted that Havilah was located just across the Red Sea from Ophir. That is very important because if Havilah still existed in name and place, then it would share a common border with Ethiopia, another clue mentioned in the Bible.

So, my hypothesis is this; as I previously stated, I believe that the northern border of the Garden was located in southern Turkey, just south of Gobekli Tepe. But since the Bible includes places with 'gold', and even Ethiopia, that means that the

Garden was not just a small place located near the Persian Gulf. It means that it must have also included present day Lebanon, Syria, Israel, Iraq, Iran, Jordan, most of Saudi Arabia and part of Yemen, and most all of Egypt, Sudan, and Ethiopia.

In other words, the Garden of Eden was a very large place that could easily feed and provide for millions of people inside of it, as proven by the populations of the present-day countries now located where it once existed.

So, what about the other two rivers mentioned in the scriptures?

Well, I believe that the Nile River, which actually flows from Ethiopia, where it's headwaters begin, on down through present-day Sudan, and then on into Egypt, was another one of the rivers that watered the Garden.

And so, very probably, the last river to water the Garden was the Jordan River, with the Sea of Galilee included in it, and along with the present day 'Dead Sea' having been a huge, fresh water lake at that time, existing right in the middle of the Garden.

So, again, the Garden of Eden was a huge

place, encompassing most all of the present-day Middle East, and northern Africa. Now, let me give you some more reasons why I say the Garden of Eden was located in this huge area.

I also believe that the Garden was located here because when God called Abraham out of Haran (southern Iraq) and brought him to present day Israel, God promised to give to Abraham; *Genesis 15:18-21 In the same day the LORD made a covenant with Abram, saying, Unto thy seed have I given this land, <u>from the river of Egypt unto the great river, the river Euphrates:</u> The Kenites, and the Kenizzites, and the Kadmonites, And the Hittites, and the Perizzites, and the Rephaims, And the Amorites, and the Canaanites, and the Girgashites, and the Jebusites.*

Scripture plainly says that God gave the land from the Nile River all the way up to the Euphrates River to Abraham. And if you research where all of the Tribes mentioned in these scriptures were once located, you will find that they occupied much of the same area where God's little slice of Heaven on Earth was once located before they occupied it.

You may ask why God was going to destroy all of the inhabitants of the lands mentioned above and give those lands to Abraham and his seed.

Well, the answer is very simple. All of these races of people had been contaminated, and *inseminated,* by the so-called 'sons-of-god', and they were full of 'giants' living among them! In other words, they were full of evil, satanic replicas, of himself!

But let's look at another reason of why this large area is where the Garden of Eden once existed. If you take a map of the world and place your finger at the landmass that basically sits directly in the very center of it, you will be looking at exactly where the Garden of Eden once existed. This location in the very center of the Earth was, and still is, very important to God.

"Why?" You ask.

Well, do you remember when I mentioned about the light from the Big Bang exploded outwards *in all directions* at the beginning of Creation?

I also said that would have placed Heaven directly in the very center of the Universe, with

God being completely surrounded by all of His creation on all sides. So, it only makes sense that God would want His little slice of Heaven on Earth to be located directly in the center of the Earth, too, just like His own home is in the Universe, surrounded by all the spiritually 'good' people of the Earth, once Satan's replicas had been *'subdued'* and God's people had *'replenished'* it.

We could, also, if we were to take the time, get into the scriptures where God calls Jerusalem His eternal home, along with others that declares Jerusalem will be the place on Earth where Jesus will set up His thousand-year reign as King over the Earth. So, again, look on a map and see just where Jerusalem is located. You will find that it sets right dead center in the very center of the Earth!

So, don't you see, all of that land had once belonged to God's first creation of Mankind, Adam and Eve, and God was trying to take possession of it back and then giving it to another of His righteous people through Abraham in another attempt to set aside a little slice of Heaven here on Earth!

*This is expanded on, plus many other instances of where God has tried to establish specific areas on earth for His righteous people, in "The Never Ending War", another book I have written.*

Also, if you use just plain commonsense, you have to realize that for Adam, Eve, and their offspring, to produce enough progeny to be able to go out and "subdue" the entire world, and then to be able to "replenish" it, they would, by necessity, had to have a huge swath of land to live in. And, the land would, by necessity, had to have been fertile enough to support and feed them all. In other words, the land would have had to support hundreds of thousands, if not millions, of people!

Another thing I find absolutely fascinating is that most of the land area described above that was once covered by a beautiful, fertile Garden, is now covered in layers and layers of sand; a huge desert! But Climatologists say that around 10,500 years ago, the Sahara Desert, which covers parts of Ethiopia, Sudan, and Egypt, was once covered in a huge rain-forest type environment.

Need I say more?

Okay, thanks. I will.

After Satan's man took possession of the land

because of the sins of Adam and Eve, God used *'the laws of nature'* to cause the beautiful, bountiful Garden to die out, and then He caused *'the laws of nature'* to totally cover over and bury all of the vast organic matter that once existed there with dirt and sand.

I'm sure that Satan was completely devastated when God did this. Because, I'm sure that the Garden reminded Satan of the 'home' he had gotten kicked out of; Heaven. I'm sure he was really homesick, and once he deceived Adam and Eve, he thought he would be able to take possession of and enjoy the beauty and comforts of the Garden, since it was an earthly replica of what his old home was.

It didn't happen, though. God sent him packing once again with his draconian tail tucked securely between his legs by turning the Garden into a huge rain-starved barren desert!

But watch what then happened when God again turned the evil Satan had caused into *good.*

God, again, turned what Satan had meant for evil by deceiving Adam and Eve, and then turned it all into good. Because, you see, buried deeply under most of those layers and layers of desert sand was all the millions, or billions, of tons of organic remains from the Garden of Eden. And instead of God letting it lie there and rot away, He turned it into the world's largest deposits of oil. Now, for that to have happened, it would have taken tons and tons and tons of *'organic'* matter to have been buried deep under the present-day soil where it would have been compacted and eventually turned into oil.

So, the question must be asked; where did all of the 'organic' matter come from?

Ah, probably from where the Garden of Eden had once been located! Me thinks so!

Now, my final reason for believing this entire area was all once the Garden of Eden is because of the following scripture: ***Isaiah 35:1-2*** *The wilderness and the desert will be glad, and the Arabah will rejoice and bloom like the crocus.* ***It will blossom profusely and rejoice with rejoicing and shout of joy. The glory of Lebanon will be given to it, the majesty of Carmel and Sharon.***

This scripture says that God is going to make the Arabah (desert) *'blossom' profusely.* Now why would He do that?

It's because He was going to return that barren desert back to the glory it once had! And if you go to Israel today, you will see exactly where this scripture has been fulfilled! The Arabah desert blooms today in full majestic Godly glory!

Okay, this finishes my expose on the Garden of Eden. So, what about Cain and Abel?

Well, we already know that Cain was deceived by Satan, and as evil overtook his heart, in a fit of anger and jealously, he killed Abel. Following that, God sent him away. But once he left, the Bible says he got married, and then eventually built a city. He also complained to God that everywhere he would go that 'people' would try to kill him. So, God put a mark on him to protect him from these 'people'.

So, who were *these people* that may attempt to harm him? And what about his wife? And what about the people who lived in the city that he eventually built? Where did all of these 'people' come from since the Bible had only named Adam, Eve, Cain, and Abel at this time? So, if you read the Bible absolutely true as literally written, there should have been only four people on the entire Earth at that time.

So, what is going on here?

Well, the answer is simple. Since we have to assume that these 'people' Cain was afraid of that may do harm to him, had to have evil in their hearts, thus, they had to consist of the evil offspring of Satan.

This has to be absolutely factual since Abel was now dead, and Cain certainly would not have been too afraid of Adam and Eve, that is unless Cain was going to try to take Eve as his wife since she was the only female listed in the Bible at that time. I would imagine that Adam would be very angry at Cain if he would have attempted that.

But that leaves us to wonder who the 'people' were that occupied the city Cain built.

Again, they had to be offspring of Satan. Who else would have wanted to live near a convicted murderer other than other evil people?

So, without a doubt, all of this leaves only one true answer. Satan had brought forth upon the

Earth what can only be called pre- and current, Adamic human lookalikes. Proof of that exists if you look no further than some of the inmates that occupy the prisons of our societies.

Listen, don't be so shocked! Listen closely to what I'm saying here! God created the 'human race' through His direct creation of Adam and Eve. But unless some of Adam and Eve's offspring that they produced while still living in the Garden of Eden had an encounter with Satan and sinned before Adam and Eve did, and that episode is just not included in the Bible, then that leaves only one other solution. And that is that Satan brought forth his own evil imitations of God's good creations out of the natural laws that God had included on the Earth when God had caused it to be formed. In other words, Satan caused his own form of life to appear upon the Earth in a very closely resembling form of God's own creation.

But I just have to ask this question; do you think that there is a possibility that the city that Cain built is still here today?

Well, it's just possible it could be. There is a place in southern Turkey just outside the northern border of where the Euphrates River begins, and where the probable northern boundary of the Garden of Eden was. It's a pre-historic ruin called, Gobekli Tepe, that I mentioned earlier.

Gobekli Tepe dates back to over twelve-thousand years. Back almost to the time that the Sahara Desert was still a rain-forest. But learned scholars state that it should have been impossible for 'mankind' to have built such structures at that *'yom'* because according to them, mankind was still just in the 'hunters and gatherers' phase of history. (Guess that tells you just how 'learned' these scholars are)

But yet, huge, multi-ton stones were carved and smoothed, and then somehow stacked on top of each other in precise alignments with each other. This was supposedly at a *"yom"* when mankind barely knew how to make stone tools, at least according to the so-called experts.

But yet, this pre-historic site also has carvings etched into the stones depicting all kinds of animals, many of which were definitely never found in southern Turkey. These prehistorical carvings are the equal to any artistic carvings ever done by any artist ever! But these carvings lead to this question. Who could have even known that some of the animals that are carved into the stones even existed? And how would they even have known what they looked like unless they had seen them before?

Take for an example, one of the carvings is of a monkey. Well, monkeys certainly are not a specie that is found in southern Turkey! In fact, the closest place where monkeys probably were found at that time was quite a distance away down in Africa, possibly around Sudan and/or Ethiopia. In other words, they would be found near the southern boundary of the Garden of Eden.

So, how did Gobekli Tepe come about? Who built it? And why?

Well, I believe I know the answer to these questions. And the answers are in the Bible.

**Genesis Chapter 4:17, says; *And Cain knew his wife; and she conceived, and bare Enoch: and <u>he builded a city</u>, and called the name of the city, after the name of his son, Enoch.***

So, based on this scripture, I truly believe that Gobekli Tepe was the so-called city that Cain built. By Cain being the founder of Gobekli Tepe, it is completely plausible, even probable, that he would have been completely familiar with animals such as monkeys that are carved upon the walls and forms of the city. I am sure that Adam and Eve passed on their intimate knowledge of all of the wonders and amazing creatures that existed inside of the Garden to their children. Remember, Adam had named many of these animals and creatures!

And the final reason I believe it could be Cain's city is that God left it here for our benefit, to show us how if we don't maintain a 'good' and righteous relationship with Him, that we will be on the 'outside' looking in!

## CHAPTER ELEVEN

## "Final Proof for Existence of God"

Even though the proof for the existence of God

has already been proven in this book, I know that there may still be some out there who are not yet fully convinced. So, I am going to present my final argument.

I truly believe that the only reason why some people will not accept the 'fact' that there really is a Supreme Being most of us call God, is because they can't actually see Him. Even though you have seen with your own eyes, and know in your heart, the astronomical 'miracles' presented here that could not have happened in any other manner other than by a Supreme God, some of you will still not believe in Him because He is an invisible mystical figure to you.

Well, I'm going to try to change your mind just one more time. And the only way I can think of to do that is to go back into the vastness of the astronomical Universe for that proof.

So, I want you to go outside tonight when it gets dark and look up into that beautiful, glorious sky that is inexplicably and unexplainably filled with 'matter'. You will see that particular matter in the form of stars, moons, and other planets in our solar system.

Now, you know all of those things exist because you can 'see' them.

But you know what?

As you are looking up into the Universe, even though it looks filled to the brim full of sparkling, twinkling matter, what you are *not seeing* is around 80% of the true real amount of matter that actually makes up the Universe.

That's right!

Around 80% percent of all of the actual matter in the Universe is what is called, "Dark Matter'. We know it exists! The proof is in the astronomical gravitational pulls, pushes, and gyrations it interacts with on the other heavenly bodies that we can 'see' with our own eyes.

So, we know, without any shadow of any

doubt, that "Black Matter" exists. Physics proves its existence! Mathematics proves its existence! So, only a fool would doubt its existence.

Now, if you have a problem believing this, then all I can say is that you're a hopeless case. But if you do believe that invisible dark matter really does exist, even though you can't actually see it, then I must ask why you still have any argument at all to now believe that there is a real Supreme, but invisible, God that created it all.

You know, there are many astrophysicists, astronomers, and quantum physics scientists, that if you asked them if they believe in a God, most of them will say something to this effect; *Intellectually, I must believe that a God exists. There are just too many unexplainable things that happened for the Universe to exist for there not to have been a Supreme Being that created it all.*

*But emotionally, I wish there wasn't a God because even with all of my education, along with all of the other smart, educated people in the world combined, still cannot explain the 'miracle' of Creation in any other way than the 'fact' that an* **invisible** *Supreme God created it all!*

But, my friends, *'you'* are not these

astrophysicists, astronomers, and quantum physics scientists. At least, most of you aren't. So, let me bring this down to a very personal level for you.

Not only does physics 'prove' that a *Superior Entity* of massively greater intelligence than any human being created Creation, the proof is in the pudding, as they say. Again, the *'natural law of physics'* proves this!

But not only does the *'natural law of physics'* prove this is true, the undeniable proof of the *'natural law of mathematics'* also proves, without any shadow of a doubt, that a *Superior Entity* of massively greater intelligence than any human being created Creation.

But still, even though all of this 'proves' the existence for 'God', it still doesn't feel right to many of you. So, again, let me bring this down to a very personal level for you.

Somewhere around 3500 hundred years ago, a man named Moses had a very personal encounter with this 'God'. This God was telling Moses He wanted him to return back to Egypt and then bring all of the Hebrew people out of Egypt.

Eventually, after much back and forth, and after Moses finally got over all of his fear and whining, Moses finally asked this God 'who' could he say to the people that it was that told him to do this.

God replied, "Tell them **_I AM_** sent you!"

"*I AM?* What kind of a name is that", most of you are saying.

Well, let me tell you.

**_I AM_** means that God is not a "I WAS"!

It also means that God is not a "I WILL BE"!

His name means exactly what he said it was; **_"I AM"!_**

That's because He is the exact same God 'now', at this present time, as He was when he spoke to Moses. He even inspired a man name Malachi to write the following and for it to be put into the Bible: **_Malachi 3:6 God is always the same yesterday, today and tomorrow. He never changes. He is so faithful, so constant, so loving and so true!_**

God is the same Being today, *("I AM")* that He was 13.799-billion years ago when He spoke Creation into existence!

This means, that no matter where you are, or what you're doing, at any 'time' or any 'place', the great *I AM* will meet you there at that time and place. All you have to do is call to Him.

Just say, "God, even though this book makes a very good argument for your existence, I'm still not 100% percent convinced yet. But I really want to be! If you really do exist, I want to meet with you, and have a personal relationship with you. So, please God, if you really do exist, meet with me right here, right now!"

If you will say these words, or any other words you want to say, God *'will'* meet with you, *right then, right there!* And I promise that if will do this, you will never again have any doubt of God's existence!

How do I know this is all true?

Well, its like this. God gave you a mind to think. But He also gave you a heart to *'know'!*

Take 'love' for existence. Your mind may, or may not, want to love someone. But, in the final analysis, it's not your mind that has the final say about it. It's your heart. And, because of that, you'll follow the desires of your heart through hell or high water because it tells you to do so, even if your mind is screaming, *'NO"!*

Listen, after you have a personal meeting with God, your natural mind may still want to have doubts and may even actually be in some turmoil. But your ***heart*** is going to tell you that you just had the greatest, most wonderful and exciting experience you could ever, ever, ever imagine!

May the ***God*** of all Creation be with all of you! *I pray!*

www.ingramcontent.com/pod-product-compliance
Lightning Source LLC
Chambersburg PA
CBHW071258220526
45468CB00001B/181